IEE MONOGRAPH SERIES 10

Power distribution in industrial installations

Power distribution in industrial installations

T. B. ROLLS, B.A., M.A.M.E.M.E., M.I.&S.Inst., M.A.I.S.E., M.Cons.E., C.Eng., F.I.E.E.

Consultant, McLellan & Partners, London SW1

PETER PEREGRINUS LTD. 1972

Published on behalf of the

INSTITUTION OF ELECTRICAL ENGINEERS

Published by Peter Peregrinus Ltd.
2 Savoy Hill, London WC2R 0BP

ISBN: 0 901223 14 X

Printed in England by
Hindson Print Group, Newcastle upon Tyne

PREFACE

The requirements of electrical distribution in industry must vary to suit the needs of the factory or works concerned, with the result that no two installations are the same. The underlying engineering principles do not differ greatly from those relating to public supply systems, but the applications of these principles are by no means the same.

It is the hope of the author that this monograph will help to direct the engineer, faced with designing a distribution system, towards the problems which must be solved to achieve a sound and economical installation. Many subjects are involved, and it would probably be true to say that nearly every chapter could be expanded into a separate book, so that some omissions are inevitable. In particular, it has not been possible to deal with commissioning and maintenance.

During the preparation of this book, a wind of change has been blowing.

(*a*) Economic considerations have brought about many amalgamations in the electrical manufacturing industry, and, in their wake, discontinuation of many old lines. In the short term, this causes the user some inconvenience with respect to spares and when making extensions. Looking further ahead, as long as the choice of equipment is not too restricted and what is available represents an improvement on what was previously available, the user should ultimately be able to share in the benefits of the reorganisation.

(*b*) In the past, British Standard specifications have dominated the construction of electrical equipment. In their turn, many British Standards have been, or are in the process of being, revised in metric form and to conform as far as possible with the recommendations of the International Electrotechnical Commission. The user of equipment is advised to study new Standards as and when they become available. It is not practicable to summarise the changes that are taking place, except to say that there is a tendency for more detail to be included and more checks to ensure compliance, and, on the other hand, in some directions, towards working the equipment harder by allowing greater temperature increases, which may not always be appropriate for industrial use.

London T. B. ROLLS

ACKNOWLEDGMENT

Figs. 1–10 were prepared by R. H. Rockliffe (now Deputy Chairman of the North Eastern Electricity Board, lately of the Midlands Electricity Board) for a supplement to the *Electrical Review* of the 18th April 1969, and are reproduced by courtesy of the *Electrical Review* and the Midlands Electricity Board.

CONTENTS

1 Assessment of load 1

2 Incoming supplies 5

3 Design 11

4 Electrical-distribution equipment 21
- 4.1 Switchgear 21
- 4.2 Fuses 22
- 4.3 Transformers 23
- 4.4 D.C. equipment 25

5 Supply irregularities 26
- 5.1 Interruptions 26
- 5.2 Voltage dips of short duration 28
- 5.3 Voltage irregularities 30
- 5.4 Frequency variations 30
- 5.5 Vital services 31
- 5.6 References 34

6 Provision of firm-capacity heavy-current medium-voltage supplies 35

7 Protection 40

8 Earthing 47
- 8.1 Legal requirements 47
- 8.2 Technical requirements 47
- 8.3 Earth-continuity conductors 51

9 System earthing 53
- 9.1 Generator neutral earthing 54

10 Short circuits 56

11 Generator problems 58

12 Effect of power factor 62

13 Cable problems 68

14 Instrumentation 71
14.1 General 71
14.2 Ammeters 71
14.3 Voltmeters 72
14.4 Kilowatt meters 72
14.5 Synchroscopes and frequency meters 73
14.6 Kilovar meters and power-factor indicators 73
14.7 Kilowatt-hour meters 74

15 Safety 75
15.1 Legal requirements 75
15.2 Labels and notices 76
15.3 Making apparatus safe for work 79
15.4 Conclusion 86

16 Substation layout 87

17 Conclusion 90

18 Index 91

1 ASSESSMENT OF LOAD

Before a start can be made with the design of any distribution system or even the arrangements for supplying it, a knowledge of the plant load is needed, together with the details of its characteristics.

In all probability, only general information will be available in the early stages of any project, but it is then that decisions have to be made on how the necessary supply of electricity will be provided. An overall assessment of the load is essential for this purpose and, if possible, this can be done by building up the load requirements for the various areas and departments of the works.

Ideally, the installed load should be listed and, by studying the duty cycles involved, suitable diversity factors can be applied. Although information may be available in the early stages of a project for a few major items, it is unlikely that a detailed list can be completed until later on. Thus, in the early stages, approximations may have to be made that call on all the experiences and judgment available to the designer.

A useful check to this appraisal of the load is a knowledge of the electricity consumption (kWh) per unit (e.g. ton) of product produced, and even more useful are such data for each part of the process. With some information on the proposed method of operation, such as the number of operating shifts, the load factor can be assessed and the demand deduced.

For some industries, there is published information giving the demand related to floor area.*

Very large differences in the estimate of a demand can arise from different methods of assessment, and it is essential that every possible check should be made to obtain an answer that is as accurate as possible. Diversity, particularly in big works, is invariably found to be greater than might be expected even with much continuously running plant.

A major difficulty is to determine what provision should be made for load growth. A discounted cash-flow calculation will usually make it clear that no substantial amount of capital should be tied up to meet future expansion. Even if it involves some guesses, it is worth while to make at least a rough layout of the system to deal with the ultimate development of the site as far as can reasonably be foreseen; then everything not essential to the initial development can be left out. The remainder can go in as and when the various stages of the development materialise. Thereby the capital tied up in the distribution system, which is in any case unproductive, can be kept in line with the current needs of the plant.

*ALDERSEY-WILLIAMS, A. G. (1961): 'Electrical supply and distribution', Factory Building Studies 10 (HMSO)

The steps by which the capacity of the incoming supply and the various parts of the distribution system have to be increased may have to be fairly substantial. A decision on the size of unit that will allow the available capacity to be advanced in steps to suit growth requires careful consideration.

For instance, if 500 kVA transformers should prove to be convenient for providing medium-voltage supplies, further such units, together with their associated switchgear, can either be added to existing substations or alternatively located in any new substation that may be necessary to supply the additional load.

There is no need to follow such a scheme slavishly. If the load were to grow more than was expected, there would be no objection to changing to a larger size of transformer, e.g. 750 or 1000 kVA, for the later developments. It would, however, be prudent to avoid too many different sizes, as interchangeability can be of considerable assistance in emergencies. It is usually easier to install an additional transformer than to replace a small one by a larger unit, as the latter may involve a difficult alteration to the medium-voltage switchgear.

Although the preliminary design of the distribution system may have to be undertaken before full information is available, a careful record of the details must be kept of the installed load as and when they become available. This is best done in tabular form, using a separate sheet for each distribution board. Where loads of different power factor are involved, the power and reactive power should be segregated and added separately. A more accurate result can sometimes be obtained by applying different diversity factors to the power and reactive power. For instance, with a motor that runs for long periods lightly loaded, there may be little diversity with respect to reactive power but considerable diversity with respect to power.

The load on each distribution board having been assessed, it is possible to determine a suitable size for the feeder cable from the substation. Similarly, the loads obtained for each distribution board can be tabulated for each substation, together with the application of suitable diversity factors, and thereby the available supply capacity to the substation can be checked.

Again, in like manner, the loads of all substations can be tabulated and a realistic picture of the total load obtained. How much allowance should reasonably be provided as a contingency for unknowns will depend on the accuracy of the information available. Although not in quite the same category as specific allowances for future load, the effect, as has already been explained, is much the same, and any money spent on load that does not materialise is wasted.

Even when a detailed list of loads, such as the list of motors to be installed, is available, it does not follow that each motor will operate at full load under normal conditions. The reasons for this are not hard to find.

For instance, a pump will be specified to deliver so many gallons per minute against a stated head. The designer of the pump finds that he needs so many kilowatts of power to drive the pump, and includes a small tolerance to allow for any design errors. He then chooses the next larger size of standard motor. By referring to motor makers' catalogues, the standard sizes of motor are listed. If the pump maker requires 12kW, the next largest standard is 15kW, so that the pump will never demand full load. For such reasons, there is always a tendency for larger motors to be provided than are actually needed, and, where a large number of standard motors are involved, it is usually reasonable to assume that, on the average, each drive will be slightly overmotored.

Assessing the load of any group of motors will require some knowledge of how they are to be operated. By way of example, a group of three pumps might be installed initially for providing a supply of water. One pump would operate continuously, the second would operate for 60% of the time on float control, stopping and starting at fairly frequent intervals to maintain the correct level; and the third would stand by. The information relating to the efficiency and power factor of the motors can be obtained from a catalogue, from which the effective load of the three motors can easily be calculated.

If, subsequently, the demand for water increased, the installation of a fourth pump would be needed; then two would run continuously and one intermittently.

For machine tools engaged on jobbing work, there will be considerable diversity, and an appreciable part of the time, perhaps 25%, will be spent in setting up and taking down the work; even while the machine is actually revolving, it is only likely to take full load during initial roughing cuts, whereas, during finishing, which may account for much of the running time, the load on the motor may be much less. In addition, there will be further diversity where some tools are seldom used.

With other machinery such as automatic looms, there will be little diversity, and large groups of machines may operate more or less continuously at near or full load, with only a few units stopped at any time.

If the layout of the plant allows no space for the addition of further production machinery, it is reasonable to infer that there will be no appreciable growth of load in the area concerned once full production has been achieved. What may happen is that a department is retooled with more sophisticated plant powered by larger motors to obtain increased output. There is no need to install much spare capacity for such possibilities; it is better to add another transformer, or another feeder and distribution board, as appropriate, when the need actually arises.

A continuous process line, e.g. a paper mill, can be dealt with in the same way and treated as a single unit whose demand will not increase once the full production rate has been reached. Eventually the plant may become obsolete, or be subjected to a major reconstruction to cope with a larger output. When this is done, there may be a sharp increase in load; but it would be unwise to speculate on this by the provision of excess capacity; this may prove to be unsuitable for the development when it occurs, or if the new plant for the extra output is installed on another site.

It must not be inferred that future load should be ignored. This is not so. What is important is to avoid spending capital on the extra capacity until the need arises. The design must be sufficiently flexible to permit expansion with a minimum of inconvenience.

2 INCOMING SUPPLIES

The nature of the load, including the likely pattern of growth, having been assessed, suitable voltages for distribution must be decided.

In the UK, a 415/240V, 3-phase, 4-wire system has been adopted for public supply. Much standard equipment (e.g. switchgear, control gear and motors) is readily available, and no case can be made for departing from this until the magnitude of the power dictates the use of a higher voltage.

The supply authority will provide and own all the equipment up to the point of supply, which will be as near to the metering point as is practicable. Although maintaining ownership of this equipment, the supply authority may require the consumer to make a capital contribution towards the cost as part of the commercial negotiations for the supply.

In practice, the small factory or workshop will be supplied direct from the street mains or, where the load is too big for this, a substation in the immediate vicinity (Figs. 1–3). Such a substation may be interconnected with the street supply. Often the consumer's co-operation will be invoked in finding a site on some part of the works premises. While this may take up valuable space, it is to some extent an asset to have scope for additional power available on the doorstep.

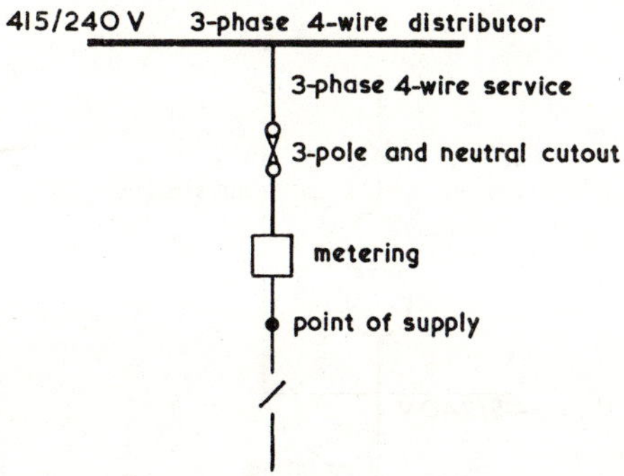

Fig. 1 Supply from medium-voltage distributor, typical load of up to 100 kVA

For larger establishments, where medium voltage would be unsuitable for handling the amount of power involved, supplies at high voltage, e.g. 11 kV or 33 kV, would be provided for the exclusive use of the consumer.

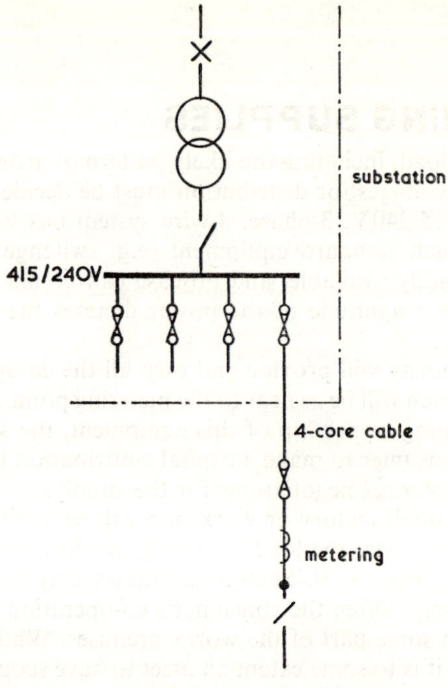

Fig. 2 Supply at medium voltage directly from substation on or near consumer's premises to typical load of 100–300kVA

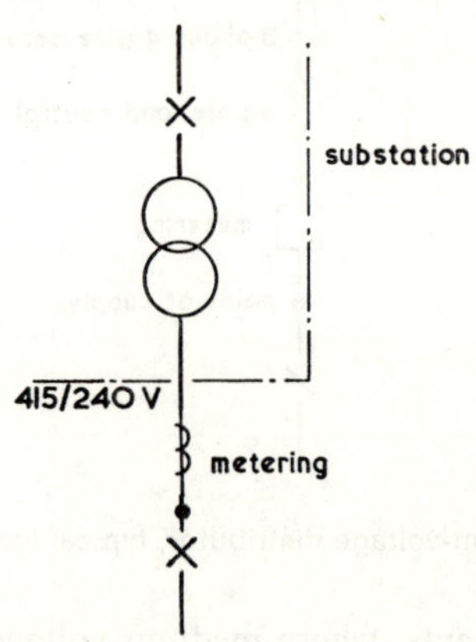

Fig. 3 Supply at medium voltage from substation on consumer's premises to typical load of 200–500kV

The consumer would be responsible for providing substation accommodation for the supply authority's equipment together with such apparatus as the consumer will himself need for controlling the internal distribution. In some instances, the supply authority may require that some, or all, of their equipment is in a separate chamber to which the consumer has no access.

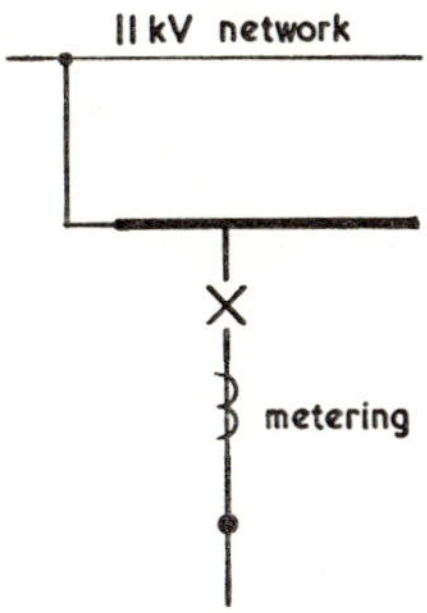

Fig. 4 Single-circuit high-voltage supply to load not exceeding 1000 kVA

Ownership and arrangement of equipment may influence the extent to which the main distribution system is subject to rates levied by the local-government authority. This is a rather complex subject; but, in general, all equipment from the point of supply up to, and including, but not beyond, any initial transformation is subject to assessment for rates. An exception is where there is no distribution board between the point of supply and the first transformer; then the first distribution board after the transformer is included in the ratable equipment.

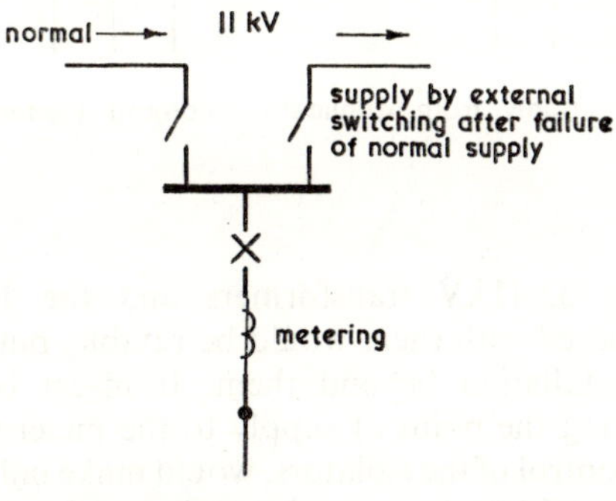

Fig. 5 One consumer's high-voltage circuit with alternative supply under supply authority's control for load normally not exceeding 1000 kVA

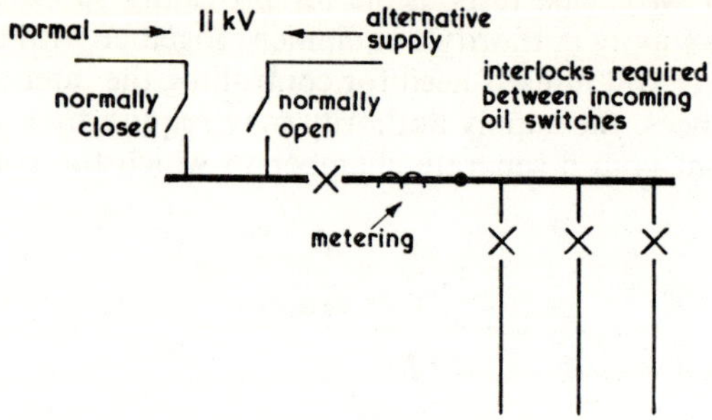

Fig. 6 Two or more consumers' high-voltage circuits without segregation for typical loads of 1000–5000 kVA

By way of example, an arrangement as shown in Fig. 7 might supply a number of substations at 11 kV. Everything up to, and including, the transformers in each substation would be subject to rates.

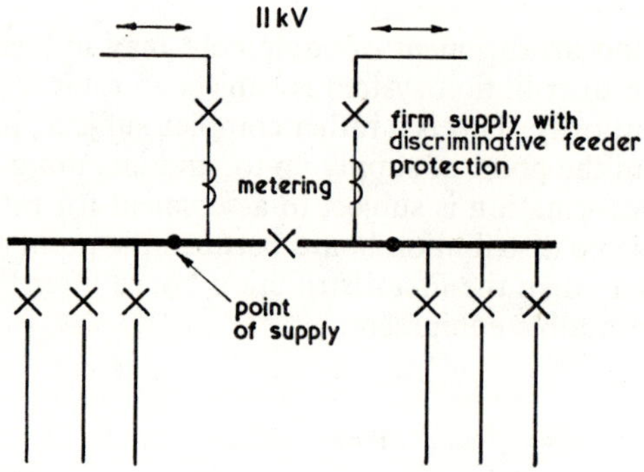

Fig. 7 High-voltage supply from duplicate incoming feeders for typical loads of 1000–8000 kVA

In Fig. 8, both 33/11 kV transformers and the 11 kV switchboard immediately associated with them would be ratable, but no other parts of the consumer's installation beyond them. It might be argued, in this instance, that moving the point of supply to the metering point, with the consumer having control of the isolators, would make only the two isolators (which would come within the general definition of a distribution board) and the two transformers ratable.

It should be noted that it is the consumer's main distribution system, as described above, that is subject to rates. This has the effect that identical equipment may be ratable in one factory (e.g. main distribution in a small works) but not in another (e.g. because it is subdistribution in a larger establishment).

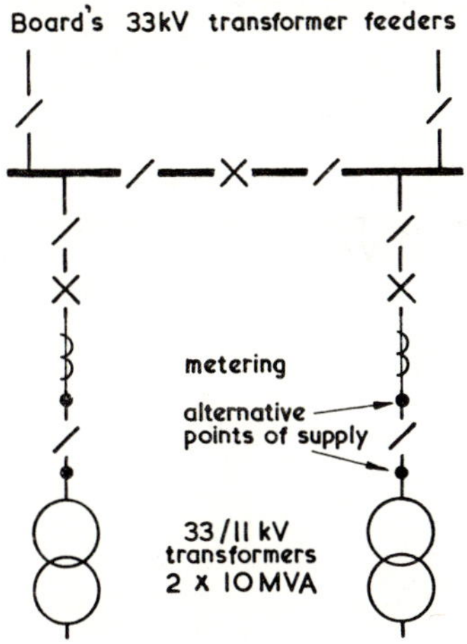

Fig. 8 Outdoor substation for extra-high-voltage supply to two consumers' transformers for typical loads of 10MVA

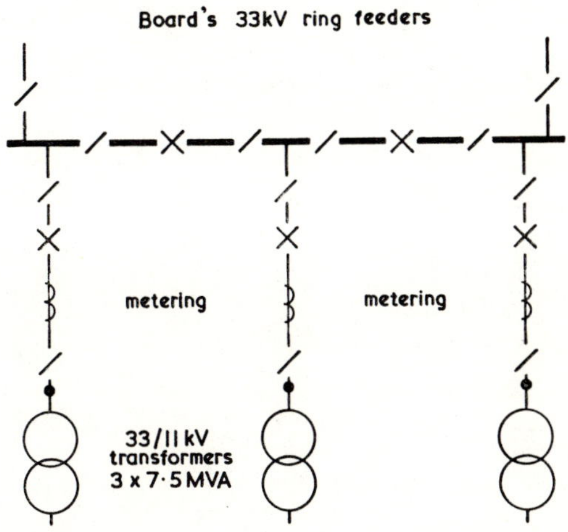

Fig. 9 Outdoor substation for extra-high-voltage supply to three consumers' transformers for typical load of 15MVA

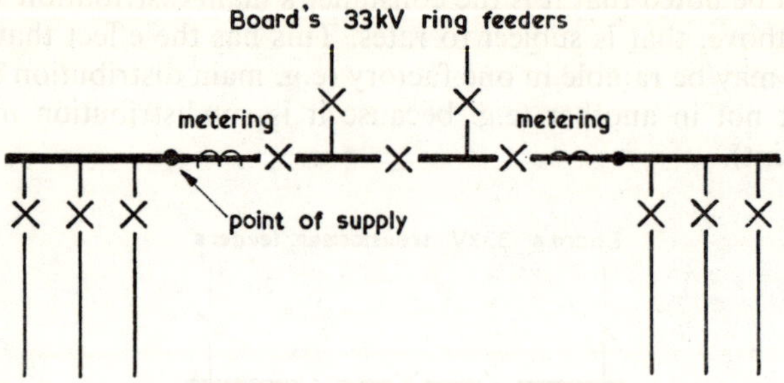

Fig. 10 Indoor substation for extra-high-voltage supply to multiple consumers circuits with typical load of 15MVA

3 DESIGN OF DISTRIBUTION SYSTEMS

Much of the design of a distribution system involves splitting the load into manageable units that can conveniently be supplied by circuits from the nearest main distribution point or substation. The size of each unit will be influenced by many considerations.

First, each unit must comprise loads that are located close together. The needs of the process may upset this requirement to some extent. For instance, if a plant includes two independent process lines, the electrical distribution should be arranged to supply each separately. It would be wrong to supply the entry end of both lines from one circuit and the delivery ends from another, so that a fault on either circuit stops both lines. In another instance, the needs of a factory might be particularly dependent on compressed air; it would be then wise to divide the compressor-house supply into two circuits, even if one circuit would be adequate for the current needed. It will be obvious from these examples that the more the electrical engineer knows about the process, the more intelligently he can apply his specialist knowledge to design the most suitable distribution system.

Secondly, the total load of each unit, after allowing for diversity, must be capable of being supplied by circuits which are reasonable in size and first cost and to which there are no other technical objections. For instance, if the load of any substation is large with respect to the voltage, various problems, such as the breaking capacity of the switchgear, provision of cable runs and voltage drop, may be encountered. It is also necessary to determine whether the 'firm' capacity (i.e. for which standby is provided) or the total installed capacity is available for use.

Whereas it is not possible to lay down hard and fast rules, it is suggested that the maximum loads that can conveniently be supplied from any substation be as follows:

Voltage	*Load, MVA*	*No. of transformers*
415	3	2 or 3
3300	7·5	2
6600	12·5	2
11 000	20	2

For the majority of installations, whether supplying the complete needs of a small factory or a part of a larger establishment, a simple radial system will fulfil most requirements. Such an arrangement is reasonable in first cost, and there need be no qualms about reliability provided that well designed equipment, suitable for the service conditions, is chosen.

For many applications, comparisons should be made between the technical merits and costs of alternative schemes. While estimating, capital costs will not present much difficulty, and it may be necessary to take running costs into consideration. Here, constant losses, such as the core losses in a transformer, are easy to evaluate. Rather more difficult is the computation of the I^2R loss involved with a feeder or a transformer. Unless a very accurate knowledge of the load is available, there is no

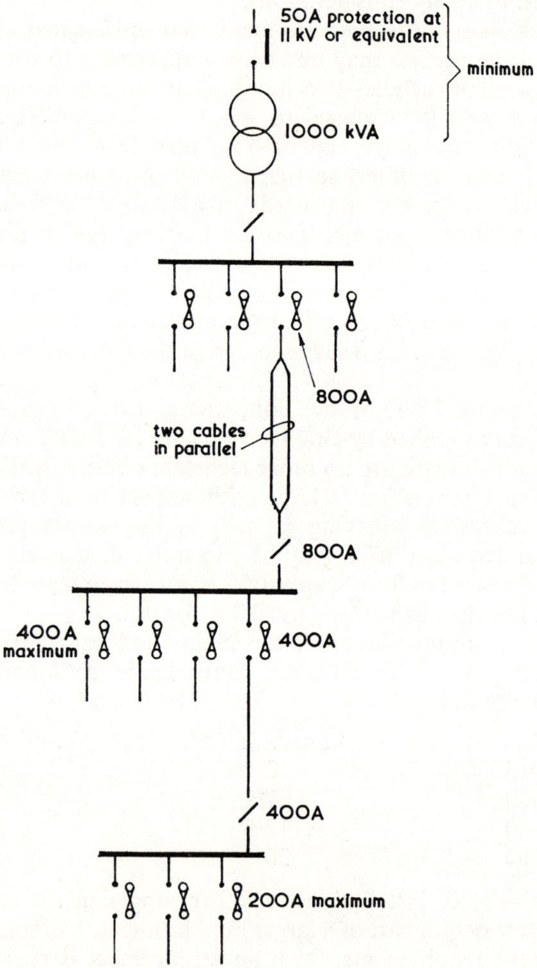

Fig. 11 Medium-voltage distribution with main, intermediate and subdistribution boards

precise way of doing this,* but the following formula provides a useful approximation:

Number of hours at maximum load = $\{0{\cdot}7\,(LF)^2+0{\cdot}3\,LF\}\,100$

where LF = load factor expressed as a percentage and divided by 10 and maximum load = the maximum applied load and not the rated full load of the circuit.

By way of example, two arrangements for supplying a given load, which is divided into two equal parts, are shown in Figs. 11 and 12. In the first

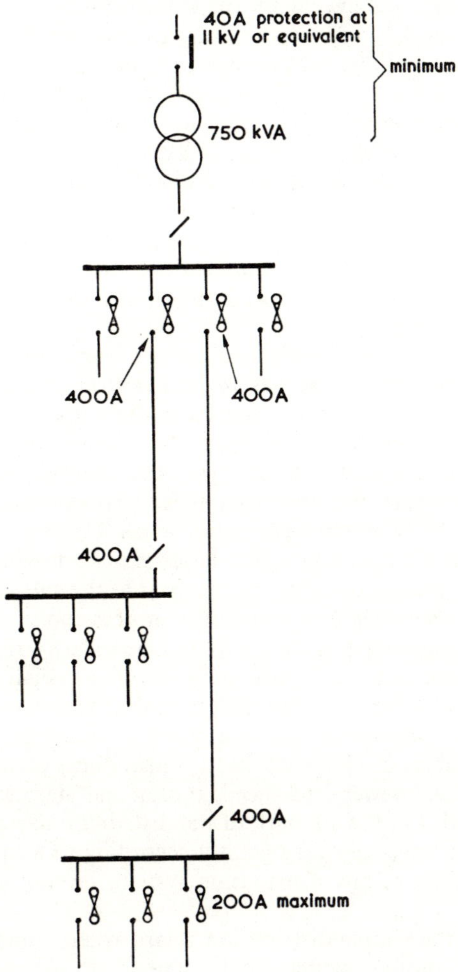

Fig. 12 Medium-voltage distribution with main and subdistribution boards only

*BEARD, J. R. (1916): 'The design of high-pressure distribution systems', *J.IEE*, **54**, pp. 125–141 (Fig. 4, p. 131)

instance, a relatively heavy feeder requiring two cables in parallel is used to supply an intermediate board and, from there, a subboard. The total load at the two distribution points is restricted by the fuses to a total of 800 A, i.e. rather under 600 kVA. Taking a single 800 A circuit from the main board implies that the incoming circuit must be protected by fuses (or equivalent) of about double, i.e. 1600 A. Such protection would be associated with that provided for a transformer of 1000 kVA or over, and would, after suitable adjustment for the turns ratio of the transformer, be located on the high-voltage side of the transformer.

On the other hand, in Fig. 12, the alternative where the maximum size of fuse is 400 A, would allow, with respect to discrimination, a transformer of only 500 kVA. This would not be acceptable for a load of 600 kVA; but, assuming that there is not much other load, and allowing for diversity, it might have been possible to use a 750 kVA transformer. The maximum size of cage motor that can be started direct-on-line will also often need consideration. In Fig. 11, the maximum size of motor, at the intermediate board, will be twice that of the subboard and twice that of either of the subboards in Fig. 12. For many applications, the arrangements shown in Fig. 12 will not only prove cheaper but will also avoid concentrating too much load in one circuit.

A similar problem sometimes arises where a single large motor, together with a number of small ones, has to be fed from the same distribution board and obtaining adequate discrimination proves awkward. It may not be possible to achieve this without the cost of undue strengthening of the system with respect to the ratings of the switchgear or the size of the cables. In this context, the need to achieve perfection in protective gear may not always be of paramount importance. The small motors may be auxiliaries to the big one, and, if the latter should develop a short circuit and operate the fuses or other protective gear at the substation, it may not matter losing the supplies to the auxiliary motors too.

A further consideration is that it is not necessarily right to make the distribution system suit the requirements of one especially large motor; it may be better, and sometimes essential, to make the motor suit what supply is available; e.g. a slipring motor or a reduced-voltage starter with a cage motor may be viable. In a large works, this introduces yet another problem, in that, for a given power and speed, motor and starter standardisation will be demanded, but the strength of the distribution system to deal with direct-on-line starting, may vary greatly according to locality.

At the extremities of any distribution system, several arrangements are possible.

First, there is the simple distribution board where, unless the supply is controlled from another switch in the immediate vicinity, an incoming isolator is required to provide convenient local means for disconnecting the supply. Although not normally needed to break the load, such a switch must be capable of interrupting the normal load current in an emergency, so that any quick break switch is suitable. The outgoing ways are sized to suit the needs of the final subcircuits, but need not all be the same size. The protection provided for each outgoing way can be in the form of either cartridge fuses or circuit breakers complying, respectively, with BS88 or BS3871.

Secondly, there are more comprehensive arrangements of distribution boards built up of a number of fuse-switch units, each controlling a separate circuit, to form complete cubicles. The result is a neat and compact arrangement; but the cabling must be carefully laid out, as the space available is often restricted, and it is not always possible to connect up a spare way without isolating the complete board. A variation of such an arrangement is to incorporate motor control gear in the outgoing ways, so that the distribution board becomes wholly, or in part, a motor control centre. Where motor control gear is involved, the requirements of BS4070 for high prospective fault currents will apply. Although it is not strictly applicable to installations without motor control gear, it would appear consistent to adopt at least the principles of this Standard to all panels of similar construction where only fuse-switches are involved. Rather less ambitious, but more suitable for making up on site and with flexibility to cope with varying requirements during the life of the installation, is the provision of a busbar chamber with fuse switches mounted above or below as may be required.

The last arrangement to be discussed is the overhead busbar system. This comprises a continuous run of busbars with facilities for plugging in fuse units at frequent intervals. It is viable where it is required to leave the electrical distribution intact during retooling of a department, the need for which may occur at relatively frequent intervals in some industries. There are some limitations. First, the density of the load must be high to ensure good utilisation of the facilities provided. Next, the size of the majority of the motors must be such that they are suitable for use with plug-in units of 30 or 60 A, with an upper limit of 100 A. The overhead busbar scheme may prove unduly extravagant where a multiplicity of small motors has to be supplied, i.e. where 30 A plug-in units would be too big. The same applies where most of the load comprises a relatively small number of more widely spaced larger motors—such as would require 100 A plug-in units. The application of this scheme is obviously restricted where there are overhead travelling cranes. The need to use a ladder and to provide flexibility in the cabling for withdrawing the plug-in units is sometimes considered a disadvantage.

Reverting to the larger installation where a high-voltage incomer is provided by the supply authority, the density and location of the load will influence the distribution arrangements, and it may be necessary to compare several schemes.

In making comparisons on an economic basis, it is important to compare schemes which provide comparable supply capacity and comparable standards of reliability. In this respect, it is not always easy to make a fair comparison.

For instance, as a matter of policy, it may be decided that standby is unjustifiable. Here, one substation at the point of supply could be used with the subsequent distribution through medium-voltage radial circuits. In the event of a transformer failure, a large proportion of the load, and in particular that which is most important at the time of the outage, can still be supplied. A fault on an m.v. feeder will affect only a part of the load, and will be more amenable to a temporary repair.

Alternatively, the substation might be moved nearer the load centre, so that the whole premises depended on a single h.v. cable. Although this comes within the parameters of what has to be accepted, a fault on this cable will cause a complete shutdown until the trouble has been located and repaired. The extra cost and losses of the first scheme might be thought to justify some extra expenditure. A detailed comparison would involve assessing the extra cost of the more reliable scheme against the probability of failure and the cost of the ensuing loss of production during the expected life of the plant. As far as the author is aware, no reliable statistics are available for the various items operating under the widely varying conditions applicable to industry, so that the need (or otherwise) to incur extra expenditure must be largely a matter of judgment.

Where h.v. distribution, e.g. 11 kV, is used within a factory, various arrangements are possible. What is best for any particular instance will be influenced by the magnitude, location and importance of the loads. The eventual choice will often lie somewhere between the extremes of 'having all the eggs in one basket' and the provision of completely 'firm' supplies everywhere. This is not easy, and again economic studies may be necessary in deciding the relative merits of possible schemes.

A simple ring main with ring-main units at each tapping point has much to commend it, and this is particularly applicable where the load is scattered over a large site. Only simple switchgear is involved, and, in the event of a fault, the affected section can be quickly found. All supplies can then be restored, after which the point of damage can be located and repaired. A fault in a transformer, or its connections, will cause only a local interruption. The likelihood of such a fault should be small and, provided that a spare transformer is available, no lengthy stoppage need be envisaged. Fig. 13 applies.

As an alternative, feeder–transformer circuits have many applications, and particularly on a compact site with a dense load. If they are used in pairs, a good measure of flexibility is obtained, and, depending on the size of transformer chosen, it will be possible to supply anything from 50% to 100% of the load if one circuit is lost.

Feeder–transformer circuits need not be operated in pairs if the degree of reliability by single circuits provides an acceptable standard.

In some instances, it may be possible to start a distribution system with a single transformer at the point of the incoming supply and from here to run two feeder–transformer circuits to different points on the site. If the load grows, the two initial circuits can be developed into a ring main perhaps with the addition of further substations. If such an arrangement appears likely to develop, some addition may have to be made to the initial conductor cross-section of the feeders. But, if the initial size had to be chosen from short-circuit considerations rather than the load of a single

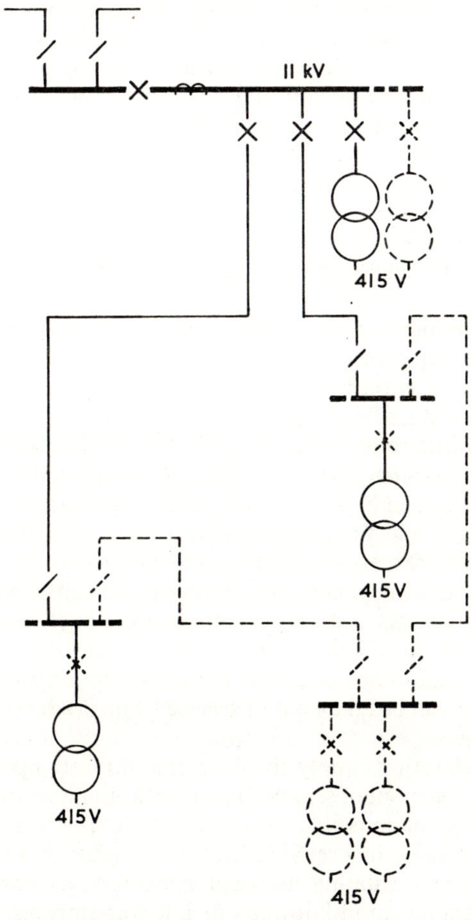

Fig. 13 Growing medium-voltage-distribution system
——— initial installation incorporating feeder-transformer circuits
.... subsequent extension to form ring main

transformer, little, if any, extra expenditure may be involved at the beginning. This principle is also indicated in Fig. 13.

One perhaps obvious point about a feeder–transformer circuit is that the transformer must be considered as part of the feeder, and not of the substation at the receiving end; i.e. it will be isolated on the h.v. side by the circuit breaker at the supply end. The transformer must, therefore, be located in a fireproof pen which is separate from the rest of the substation and for which the means of isolation will be the incoming m.v. switch supplied from the transformer. A further point is that, unless the extra expenditure of the link-type h.v.-cable box is incurred, fault location will involve breaking down the box. Perhaps here, if the system is likely to develop into a ring main, an h.v. isolator could be provided initially at the receiving end as an alternative to a link-type box. The isolator would then form part of the ring-main unit when the system is extended.

In very large installations, complicated switchgear layouts with double busbars and/or bus-section switches may be inevitable. Not only are such arrangements costly, but a greater degree of skill is needed to ensure safe operation. There is, therefore, much to be said for the simple installation that is easy to operate and where there is less to go wrong and less to maintain. Some particular aspects of this problem are discussed separately in Chapter 5.

However, an important consideration must be to match the distribution system to the needs of the plant supplied. For instance, a single process line, as referred to earlier, would be satisfactorily served from a simple switchboard. Compared with motors and control gear associated with the line, there is so little to go wrong and so little to maintain that it would be hard to justify anything more complicated. A bus section would serve no useful purpose, as one half of the line cannot operate without the other. Double busbars would provide access to one set of busbars in turn, but not to the circuit breakers controlling the supplies to the motors; and, as the busbars will need negligible maintenance, very little will be gained.

The position changes if there are two process lines. It can be assumed that each line will be taken out of service in turn when maintenance is needed. There would then be a case for a bus-section switch, so that half of the switchgear can be taken out of service while production is maintained by the other line.

Similar considerations apply to other continuously operating processes where units are taken out of service in turn without interrupting production. Here, it would be inadmissible for the whole supply to rely on one set of busbars that would be inaccessible without a major shutdown.

BS2613, relating to rotating electrical machinery, recommends minimum rated outputs, expressed in kilowatts or kilovolt-amperes, of 150, 300 and 750 for rated voltages of 3·3, 6·6 and 11 kV, respectively. This is a general guide only, and there is some latitude with respect to the exact point at which it is necessary to change from one voltage to another. Between 3·3 and 11 kV, the changeover point will be of the order of 1500 kW.

It will be obvious, however, that where, for instance, the voltage of the supply is 11 kV and the subdistribution is carried out at 415/240 V, there will be a no-man's land where motors of ratings, say, over 300 kW and below 75 kW, are not suitable for connection to either voltage.

Where only two or three drives are involved, it will be expedient to adopt unit transformers, with the switching carried out at 11 kV. The motor voltage is then of not much immediate significance, and could be 415 V if the transformer can be sited near the motor.

Owing to switching surges, reasonably liberal insulation of the transformer is necessary with this scheme; and, while the author is not aware of any troubles with oil-immersed transformers, some difficulties are known to have occurred with air-insulated transformers.*

With more drives, it will be necessary to introduce an intermediate system, e.g. at 3·3 kV. In general, this should be reserved for the motors to which it is applicable. While transformation of small amounts of power from 3·3 kV to 415 V may be convenient, it may be found more economical to transform directly from 11 kV to 415 V wherever practicable.

If a system is likely to grow whereby initially one or two drives with unit transformers are needed but a 3·3 kV system will be required later, it would be better from the point of view of motor standardisation to adopt 3·3 kV as the motor voltage from the start.

6·6 kV is not extensively used for new installations, but it has much to commend it where the ultimate power requirements at this voltage will not lead to excessively high currents. There is practically no gap, and the changeover point between 6·6 kV and 415 V is in the region of 300 kW.

Outside the UK, other voltages are encountered, and, on the continent of Europe, 380 V 3-phase 4-wire systems are usual. They are used in conjunction with systems where the high voltage ranges between 6·6 and 20 kV. In general, the system designer would be advised to follow the standards of the country concerned, thereby deriving the benefits of applicable standardisation.

550 V 3-phase systems are encountered in mines where voltage drop becomes a problem as working extends from the shaft. Underground, the need to provide secondary transformation for lighting is always present, but, where the use of 550 V spreads to surface installations with high lighting intensities, the inconvenience of a separate 415/240 V lighting system must be faced. 550 V systems present no difficulties where motors are involved, and the no-man's land referred to earlier before the next higher voltage, e.g. 11 kV, will be less. As much of the 415 V switchgear

*WHITE, E. L. (1964): 'Investigation of switching surges in a high-voltage motor installation'. ERA Report 5055

and motor control gear is equally suitable for 550V, there is economy of installation. While this voltage will continue to be used for mining, it is unlikely to be used extensively elsewhere and, even for mining, consideration is being given to going up to 1100V to mitigate voltage-drop problems with increased mechanisation.

Direct-current systems may be required in some works for supplies to cranes and other auxiliary plant that can conveniently be driven by series motors. Where this is done, the use of d.c. may be extended where shunt motors can be used for variable-speed drives. Opinions differ as to whether a.c. or d.c. motors should be used for cranes etc.

The provision of a d.c. system presents no insuperable difficulties, but it may be necessary for an overall comparison to be made, taking into consideration not only the distribution system but also the drives. By the use of thyristor units for individual drives, it is possible to achieve the fine control associated previously with Ward Leonard sets, but, so far as distribution is concerned, this is an a.c. problem involving both harmonics and reactive-power swings.

Where d.c. systems are required, voltages of either 220/240 or 440/480 are used.

For rectification equipment, the higher voltage range will be more economical, although this advantage has become less pronounced with silicon diodes than was the case with mercury-arc rectifiers. Despite this, the lower range is sometimes preferred on the ground of greater reliability under adverse operating conditions.

Two particular points in the design of d.c. systems must be mentioned. First, care is needed that protection on the outgoing d.c. circuits is fast enough to obviate fortuitous operation of the diode fuses. Secondly, at times of light load, regeneration can give rise to objectionable voltage increases, and some loading resistance may be needed.

Lighting supplies are sometimes derived from separate circuits or even transformers; this practice often arouses controversy. On the other hand, lighting supplies are simple, and there is a reasonable chance that they may be maintained if a failure occurs to the heavily loaded power circuits; also, the lights are not affected by voltage variations due to fluctuating loads. However, if lighting is merged with other power supplies, the cost of the installation will be less, and machinery is unlikely to continue running if the lights are out.

In either case, it is necessary to decide what degree of emergency lighting is needed. As failures can be local, it is best to provide emergency supplies that are controlled on a local basis. Except where special risks are involved, self-contained automatic units incorporating sealed batteries and chargers, located at a few strategic points, provide all that is necessary to avert panic and to allow personnel to be evacuated.

4 ELECTRICAL-DISTRIBUTION EQUIPMENT

The maximum utilisation for which any item of electrical equipment can be employed is largely dictated by the relevant British Standard. The co-ordination of the various items that go to make up a distribution system is by no means easy, as the ratings derived from British Standards are designed to suit the needs of the equipment rather than the requirements of the circuit. For instance, if a transformer circuit requires a rating of 1000 kVA, there may be no corresponding circuit breakers suitable for the high- and low-voltage sides of the transformer; it is, therefore, necessary to choose the breakers that will permit the desired flow of current with a minimum of excess capacity. As, in some instances, the cost of the switchgear and connections may even be greater than the transformer being controlled, there may sometimes be a case for choosing a transformer to suit the switchgear.

4.1 Switchgear

In addition to requiring a comprehensive set of type tests for short circuits, it is important to remember that BS116, which covers oil circuit breakers, is based on an average ambient temperature of 35°C, provided that the maximum does not exceed 40°C over a period of 24h. For the majority of applications this is adequate, but, for use at even higher ambient temperatures, some derating may be necessary; no allowance is permissible for lower ambient temperatures. The extent to which derating is necessary will depend on the margin (if any) available during type tests for temperature rise. As type-test certificates will apply to the manufacturer's standard product and as the normal current in accordance with BS116 must be stated on the rating plate, it is often expedient to give the circuit breaker an extra label giving particulars of the derating applicable to the site conditions. It will be apparent that the amount of derating will depend on how closely a design approaches the limits of temperature rise imposed by BS116, so that, in practice, a particularly liberal design may be capable of working in an abnormally high ambient temperature without the need for derating.

Circuit breakers which have to operate at altitudes in excess of 3300 ft are also subject to derating. It should be noted that circuit breakers are available only in certain standard current ratings and also standard breaking capacities. If a suitable standard is not available, it is necessary to go to the next larger size. Standard service voltages also apply, but it must be noted that, where it is necessary to go to the next higher voltage standard, the breaking current of the latter standard will apply, so that the breaking capacity expressed in megavolt-amperes at the operating voltage will be correspondingly lower.

It is worth remembering that type tests are carried out on new equipment, and no allowance is made for deterioration in service. Some items of equipment in industrial conditions may be required to carry load for weeks, or even months, on end without intermission or opportunity for maintenance. In such circumstances, some derating will be a wise precaution, especially where there is no margin with respect to ambient temperature.

While the above applies particularly to switchgear supplied in accordance with BS116, the same principles apply to all switchgear.

In choosing switchgear, it is not only the circuit breaker, whether oil- or air-filled, but the whole equipment that must be considered. Such factors as means of isolation, provision for earthing, safety interlocks and ease of maintenance are of considerable importance. These facilities are provided as a matter of course by leading manufacturers of h.v. switchgear and much m.v. switchgear too.

For some m.v. circuits, it may be possible to use miniature and moulded-case circuit breakers constructed to the requirements of BS3871, where the prospective short-circuit currents are not in excess of the ratings available. Such circuit breakers do not normally have facilities for isolation, so that it is necessary to resort to the use of a special adaptor to prevent closing the breaker when work is being done on a circuit; even so, it is not possible to see that the contacts are fully open. Usually such breakers are supplied as sealed units, so that inspection of the contacts after heavy loading or short circuit is not possible. If there is any doubt, the only safe course is then to replace the breaker; this involves a shutdown of the switchboard concerned. Despite these disadvantages, this type of circuit breaker has many applications on circuits where little maintenance is to be expected.

4.2 Fuses

The ranges of cartridge fuses provided by leading manufacturers include those complying with the requirements of BS88 (1967) for Category AC80; i.e. a prospective fault current of 80kA at 415V, and sometimes higher. In practice, it would be extremely difficult to obtain such a high value of current when using conventional distribution gear, so that good quality cartridge fuses never cause any anxiety about their ability to clear any prospective short circuit that is likely to be encountered.

Special consideration is needed when choosing cartridge fuses for protecting cage motors that are started direct-on-line. For this duty, one manufacturer offers cartridge fuses with oversize elements; e.g. a typical 30A cartridge of class Q1 with a fusing factor not exceeding 1·5 would be suitable for supplying a motor of 7·5hp* with a full-load current of about

*Doubtless equivalent metric recommendations will be available soon

11·5A at 415V under ordinary conditions. With a special motor fuse of the same physical dimensions, a 20hp motor can be supplied, although the fusing factor is much higher.

Another manufacturer offers fuses of class P with a fusing factor not exceeding 1·25 where a 30A fuse would be suitable for motors of up to 15hp.

It must be emphasised that, where motors are concerned, it is the duty of the fuse to protect against heavy short circuits and earth faults, leaving the protective devices incorporated in the motor starter to deal with overloads, single phasing and any other protection that may be needed.

4.3 Transformers

The effective rating of oil-cooled transformers constructed in accordance with BS171 depends both on the ambient temperature and the duty cycle. This is a somewhat complicated problem, and is described fully in CP1010; but Table 1, which is reproduced from this Code of Practice, indicates the wide range of variation that is applicable to type-ON transformers. From this, it will be obvious that considerable liberties can be taken with respect

Table 1

Recurrent daily loadings, for the operating conditions stated, expressed as a percentage of rated kilovolt-amperes

1		2	3	4	5	6	7	8
Cooling-air temperature		All day	Long period		Medium period		Short period	
		24h heavy load	16h heavy load	Remaining 8h light load	8h heavy load	Remaining 16h light load	3h heavy load	Remaining 21h light load
deg C	deg F							
0	32	120	125	105	130	105	150	105
5	41	115	120	100	125	100	145	100
10	50	110	115	94	120	94	140	94
15	59	105	110	88	115	88	135	88
20	68	100	105	82	110	82	130	82
25	77	94	100	76	105	76	125	76
30	86	88	94	70	100	70	120	70
35	95	82	88	64	94	64	115	64
40	104	76	82	57	88	57	110	57
45	113	70	76	49	82	49	105	49
50	122	64	70	40	76	40	100	40

to the loading of transformers, particularly if the maximum load occurs in Winter.

In addition to operating transformers strictly in accordance with CP1010, it is usually permissible to assign a higher rating to transformers for abnormal emergency conditions. For instance, the maximum rating of a continuously loaded type-ON transformer is based on a maximum winding temperature of 95°C, but, if it were intermittently loaded, winding temperatures up to 115°C would be permissible for brief periods. It would, however, be permissible to operate the same transformer continuously for the duration of an occasional emergency with a winding temperature of 115°C, although deterioration of the insulation might be expected if the transformer were to be subjected to this treatment for very long periods.

The supply authorities apply this principle where the concept of i.s.t. (integrated system transformers) is adopted, and a pair is used to supply an 11kV load of 23MVA. One transformer is then pushed to this limit if the other fails. It is assumed that the peak value of the loads is not maintained continuously throughout the day and, furthermore, that peak loads will coincide with a low ambient temperature. The figure of 23MVA was selected to make the best use of a 1200A circuit breaker at 11kV. This approach may need some modification when applied to industrial supplies, where there may be negligible valleys in the load curve at night and, furthermore, where the peak does not necessarily coincide with a low ambient temperature.

The operation of continuously loaded transformers under high ambient conditions, e.g. in the tropics, needs special attention, and, in accordance with CP1010, derating is likely to be necessary. This is especially so with unit transformers for generators and motors.

As an alternative to applying such derating where it is necessary to operate a standard design of transformer in a high ambient temperature, additional cooling surfaces can be provided which will enable the transformer to give its nominal output at the higher temperature. By way of example, if a 2MVA transformer is needed for a continuous medium-voltage load, a larger transformer of, say, 2·5MVA might carry the load satisfactorily at the high temperature, but the additional fault current might place the switchgear in jeopardy. Therefore it would be better to provide the standard 2MVA transformer with extra cooling.

Thermal limits under short-circuit conditions are specified in BS171, but performance, with respect to both the thermal and the mechanical effects of short circuits, is not normally checked. Some apprehension may be felt about the ability of distribution transformers supplied under severely competitive commercial conditions to cope with short circuits and switching surges that may be encountered in service.

As an alternative to oil-cooled transformers, either askerel- (a non-flammable fluid) or air-insulated transformers are available. Both these have applications where special fire risks are involved. As might be expected, extra costs are involved with both, so that, in such instances, studies will be necessary to decide whether oil-cooled transformers with the addition of fixed fire-fighting equipment or one of these alternatives are viable.

Against the askerel-filled transformer, it is sometimes claimed that toxic fumes, e.g. phosgene, are liberated if arcing occurs. This must be accepted; the transformer must be well ventilated so that any dangerous concentration of toxic fumes is unlikely.

Air-insulated transformers present no particular problems, provided that a supply of clean cooling air is available. Unfortunately, this seldom applies to industrial situations, and any remedial steps, such as filters or double air circuits, will increase the cost still further.

Where it is feasible to use either askerel-filled or air-insulated transformers in conjunction with switchgear containing no oil, packaged load-centre substations become a possibility; if space is scarce at ground level, such units can be located in a basement (given good ventilation) or in the roof.

4.4 D.C. equipment

As has been mentioned in Chapter 3, d.c.-distribution systems are sometimes encountered. Except in old installations, the source of power will be derived from semiconductor rectifiers, e.g. silicon diodes. As the latter are easily damaged, special high-speed fuses have been developed for their protection. Care is then needed to ensure that the device chosen for protecting the individual outgoing circuits is fast enough to discriminate with the rectifier fuses. Circuit breakers in accordance with BS862 will, in general, not be fast enough, so that either high-speed circuit breakers as used for traction purposes or cartridge fuses suitable for d.c. service, in accordance with BS88, must be used.

5 SUPPLY IRREGULARITIES

Disturbances in electricity-supply and -distribution systems take various forms. The following require special comment:

- interruptions in the supply
- voltage dips of short duration (say, up to 2s)
- voltage irregularities, e.g. flicker
- frequency variations
- vital services.

5.1 Interruptions

Interruptions to the supply of electricity may occur either on the supply authority's network or on some part of the consumer's own system. The old adage, 'the strength of a chain is its weakest link', is particularly applicable, and elaborate arrangements on the consumer's side of the meter may do little to improve reliability if similar precautions are not taken by the supply authority. Unfortunately, it is not easy to find a simple yardstick for assessing the strength of the links. It is, however, essential that the risks are recognised and, if it is felt that they cannot be tolerated, they must be weighed against the extra cost of providing increased security for all, or perhaps some particularly vital part of the consumer's supply, to which reference will be made later (p. 31).

A public supply authority must ensure that the continuity of service is such that, on the one hand, excessive complaints are avoided, and, on the other, the price is not too high. For this purpose, statistics are available, but these are applicable to the conditions under which networks of public supply authorities operate.

The engineer responsible for dealing with supplies within an industrial establishment is in much the same dilemma. In particular, he is unlikely to have much statistical information on which to base a reliable estimation of the likelihood of failure of the equipment subject to his control. Even if any figures were available in industry, it may be difficult to take into account how the equipment will be used, or abused, in any instance. When assessing the risk of failures, allowances, which must be made, will inevitably lie in the sphere where judgment must be exercised, and will include the following:

- load factor and/or duty cycle
- atmospheric conditions; e.g. temperature, humidity, sunlight etc.
- external conditions; e.g. dirt, corrosion and external abuse
- abnormal conditions, e.g. voltage surges and temporary overloading

- standards of maintenance
- age of equipment.

Considerations such as these are naturally hard to evaluate, but the consequences of any interruption have also to be considered. These are:

(*a*) the time at which a failure takes place; i.e. with single-shift working, any interruption in the middle of the night may pass unnoticed, but, with a continuous process, the ensuing dislocation will be more serious

(*b*) the length of time needed to restore the supply after a breakdown; here three broad categories must be considered:

(i) There are transient faults which are associated with supplies derived from overhead lines. If facilities are available for reclosing the circuit breaker that has tripped, the duration of the outage need not exceed a few minutes. Autoreclosing circuit breakers are applied successfully on some systems, and their use seems likely to increase. It is, however, important that reclosure should not take place before any voltage generated by any consumer's synchronous plant has had time to dissipate.

(ii) There are sustained faults on systems that have facilities for restoring supplies through alternative circuits, e.g. ring mains. Normally, such systems are operated in the 'open' condition, so that, if a cable fault occurs, the supply to several substations will be interrupted. It is then necessary to locate the faulty section and, by means of ring-main isolators, to restore the supplies. Obviously, the speed with which this can be done will vary with circumstances, but it is likely to be of the order of 1 or 2h.

(iii) There is the sustained fault that must be repaired before supply can be resumed. The nature of the fault and the facilities available will influence the time taken. A fault in a buried cable may take upward of a day to locate and repair, whereas it may be possible to deal more quickly with damage to an overhead line.

(*c*) the direct monetary value of loss of production; this will include intangibles such as whether the failure to achieve a delivery date may or may not have subsequent repercussions on business

(*d*) the cost of consequential damage to plant

(*e*) the frequency of interruptions; one long interruption can be equivalent to a series of short ones.

While there will be many variations, high voltage from the public supplies is generally derived from ring mains by one of the methods shown in Figs. 5–10. If and when there is a shutdown, the restoration of supply does not involve a prolonged interruption. From the consumer's standpoint, the degree of security of supply tends to improve with its magnitude rather than with the results of a failure.

Where some continuous processes are involved, a high degree of reliability may be needed if the consequences of a failure are not to be disastrous, e.g. the UK Atomic Energy Authority (Binns and Outram, 1958). Similar considerations apply to both chemical works and oil refineries, where failure of more than a few seconds may provoke consequential losses of hundreds of thousands of pounds before it is possible to get the process smoothly onstream again.

As a general rule, standards of reliability of supplies seldom give rise to complaint in the UK and the majority of consumers are satisfied with the service they receive.

5.2 Voltage dips of short duration

When a short circuit occurs, the voltage will be depressed until the appropriate circuit breaker or fuse operates to interrupt the short circuit. If a 'bolted' 3-phase short circuit occurs, no voltage will appear at the point of short circuit. The voltages on the supply side of the short circuit will increase towards the source of supply, depending on the voltage drops in the various impedances in the circuit between the source of supply, which might be the generating station, or often, in practice, the Grid, and the fault. On any healthy circuits fed from the immediate vicinity of the short circuit, the voltage will be zero for the duration of the fault.

Numerous factors vary the distribution of voltage as outlined above. First, arc resistance must be considered; here the effect will be appreciable on medium-voltage circuits in which resistance influences the magnitude of the fault current. On h.v. circuits, the short-circuit current will be determined almost entirely by the reactance of the circuit, and arc resistance may have little, or no, effect.

Secondly, most motors regenerate to some extent when a short circuit occurs. This effect is of very short duration, say 3 or 4cycles, where induction motors are concerned, but is longer with synchronous motors. As a consequence, a potential appears, and will slightly delay the operation of undervoltage releases.

Thirdly, short circuits are not necessarily balanced, so that a potential may appear between some line conductors and not others. In general, motors will continue operating under single- or 2-phase conditions, and, provided that the condition does not persist for more than a few seconds, no harm will be done. Under this condition, the undervoltage protection may or may not be connected to a phase suffering a severe drop in voltage, so that the effect will be of a random nature.

Lastly, the duration of short circuits varies. Examples of fast clearance times are

- within the first halfcycle for cartridge fuses and some designs of m.v. circuit breaker incorporating high-speed tripping devices (Hudspeth and Midworth, 1968)

- in 3–6 cycles for some designs of autorecloser (Money and Harris, 1968)
- 2–4 cycles for the 400 kV systems of the CEGB*
- up to 10 cycles for pilot-wire protection, instantaneous-earth-fault and high-set-overcurrent protection.

Where clearance times of the above order can be achieved, most so-called instantaneous undervoltage releases will ride through the disturbances, and there is insufficient time for motor speeds to drop appreciably.

Many systems, however, employ relays that have a definite minimum inverse-time characteristic set to provide discrimination on a time basis (sometimes assisted by a directional feature), which give longer delays, possibly as much as 2 s. Where faults occur on such systems, tripping of undervoltage releases is to be expected, although, as explained earlier, the effects may be of random character. In addition, any high-pressure mercury-vapour lamps will go out.

Inconveniences caused by stoppages on other than the faulty circuit may be countered in various ways. First, some form of time delay on the undervoltage protection is possible on important drives. Where unexpected restarting presents no particular hazard, undervoltage protection is sometimes omitted, for example, on power-station auxiliaries. Any attempt to adopt either of these practices universally may cause trouble if the dip is long enough for the motors to fall out of step. When the voltage recovers, all the motors will try to draw from the supply the heavy current corresponding to direct-on-line starting, which, if full voltage could be maintained, would be about six times full load. If this should happen, a severe reduction in voltage would occur, which, though restricting the current somewhat, would make it difficult for heavily loaded motors to recover, so that the sustained overcurrent would cause tripping of feeders and transformers on overcurrent.

Although it may not be possible to obviate entirely the unwanted operation of undervoltage protection, much can be done by careful selection and co-ordination of the protection to avoid unnecessarily long clearance times. For high-voltage systems, voltage dips due to earth faults can largely be avoided if the system neutral is earthed through a resistance.

In some complicated processes, even a very brief voltage dip can upset production as badly as a complete interruption if plant shuts down owing to the release of undervoltage coils. Some other form of evasive action such as automatic cascaded restarting may be needed to keep the effects of the disturbances within reasonable bounds.

*'The need for, and practical problems of, short clearance times', IEE Power Division discussion meeting, introduced by A. V. Dowell, T. B. Patrickson and N. W. Ellis, 7th Feb. 1969, London (unpublished)

5.3 Voltage irregularities

These fall into two categories:

(*a*) voltage depression occurring during regular operation of the plant, e.g. from the starting of a large motor. Provided that such occurrences are relatively infrequent, the result, though noticeable on the lights, will cause no inconvenience. If high-pressure mercury-vapour lamps are used, it is important that the voltage does not fall below about 15% of normal. If it does, the lamps will go out, and will not restart until they have cooled. Most motors will, however, continue to operate without distress for brief periods if the voltage does not fall below about 80% of normal.

(*b*) flicker due to random load variations (e.g. from the use of arc furnaces and welders) and cyclic variations (e.g. from winders and compressors).

In general, a supply system with ample capacity is the best antidote to voltage fluctuations. To put it another way, the minimum short-circuit capacity must be relatively high. For instance, an old rule of thumb for arc furnaces is that the minimum prospective fault apparent power should be one hundred times the rating of the furnace transformer (Ramsaur and Treweek, 1954), but this is now thought to be unduly cautious.

Most, though not all, load swings giving rise to voltage fluctuations are reactive (or 'wattless') in character. For such conditions, power-factor correction by means of synchronous motors or compensators tends to provide more correction as the voltage drops, and this effect can be accentuated by field forcing whereas static capacitors provide less correction with a drop in voltage. Little experience exists in the field of series capacitors, except for the rather more steady load of submerged-arc furnaces. Saturable reactor schemes, which greatly suppress the current variations of arc furnaces and similar loads, are available; developments with thyristor-switched capacitors appear promising.

If the distribution system cannot deal with variations in load without causing embarrassing fluctuations, or at least cannot do so except at excessive cost, attention must be given to the load. It may be necessary to use Ward Leonard or even Ilgner drives, which will cause less disturbance of the supply for a reversing drive than one supplied through a thyristor. Again, a synchronous induction motor may prove better, should the drive require more starting torque to accelerate a high-inertia load than can be obtained by using a reduced-voltage starting with a salient-pole machine.

5.4 Frequency variations

If the frequency increases by, say, 2–3 Hz, the speed of the motors will increase. If the torque of the driven load is constant, the load on the motors (in kilowatts) will increase *pro rata* with the speed, so that any motors that were fully loaded will then be overloaded. With some drives, e.g. fans, the torque may increase as the square (or some higher power) of the speed, which could lead to more severe overloading. Fortunately, periods of high frequency are of short duration, and occur only when a power system unexpectedly loses a large block of load. As soon as the speed of the generators increases, the governors counteract the increase and restore the frequency to normal within a few seconds. This will not be long enough to cause any overheating of the motors.

Low frequency, on the other hand, occurs when a power system is overloaded, and may be sustained for appreciable periods. The reduction in speed will reduce the load of the motors, though, owing to the increased magnetising current, the overall reduction in current will be marginal because of the reduced power factor. Damage to the motors is unlikely, but repercussions on the driven plant are possible. For example, a drainage pump that has only just enough capacity at normal frequency may be swamped if the low frequency persists.

When a power system is overloaded, low frequency may be accompanied by low voltage, which will increase the current, and the overall result on a motor that was initially fully loaded may be overheating.

5.5 Vital services

In industry, there are some instances where a breakdown can have most unfortunate results. A blast furnace is an example. Here, an electricity failure can cause havoc; in one instance, both the pumps and the blowing engine stopped. Lack of water permitted melting of parts of the cooling system in the vicinity of the tuyères. When some water did get through, it flowed into the furnace, where it made contact with red hot coke to produce carbon monoxide and hydrogen. This dangerous mixture found its way back along the blast pipe. Eventually, it was detonated, and it blew up the blowing-engine house.

It may not necessarily be best to prevent such a catastrophe by increasing the reliability of the electrical arrangements. There may be other ways; e.g. a diesel pump might provide an alternative water supply, or static water may be available. It will, however, be assumed that an earnest attempt is to be made to provide continuity of service to pumps providing a vital supply of water.

First, it must be assumed that two supplies of electricity are available from separate switchrooms, so that, in the event of failure of the works generator or the public supply, one or, if possible, two circuits remain available to the pump house (Fig. 14). To the pump house, two distinct circuits are provided (Fig. 15*a*). Nothing will be gained by installing a complete switchboard so that either unit can be supplied from either circuit. If anything, the provision of the additional equipment may decrease the reliability; e.g. a faulty bus-section switch could interrupt both supplies or, if double busbar gear is used, a fault on one may affect the other.

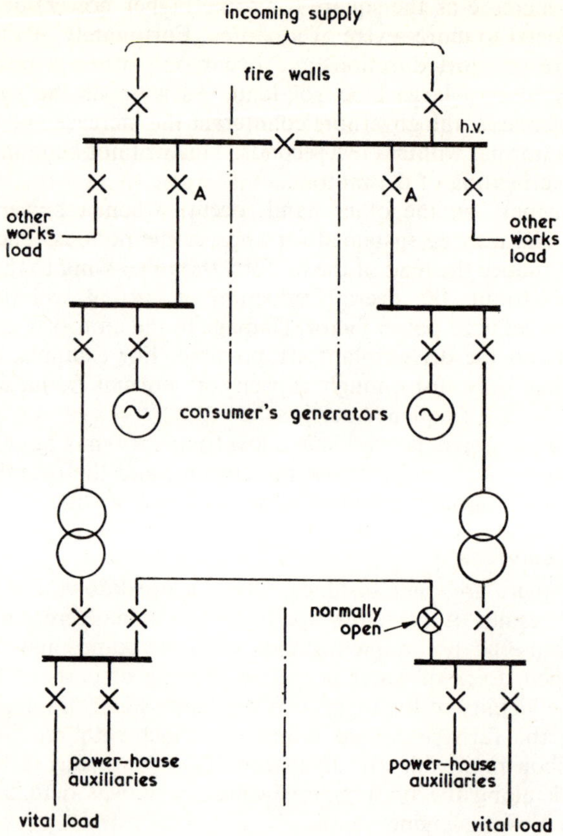

Fig. 14 Consumer's generating plant operating in parallel with public supply where circuit breakers A trip if trouble occurs on incoming supply but remain closed if works generator fails

Fig. 15*b* is a more elaborate system for three pumps. It provides for the same facility for any two pumps as previously while one unit is dismantled for maintenance.

In arranging for vital services, some sense of proportion must be retained. It would obviously be extravagant to go to extreme lengths to maintain a continuous supply of electricity if the associated mechanical equipment is not up to the same standard. A vulnerable point on some installations is the nonreturn valve. A healthy pump can waste a lot of useful water through the 'stuck' nonreturn valve of the unit that has suffered an electrical failure.

Another vital service is to provide some measure of emergency lighting (Chapter 3).

For computers, data systems and some special instrumentation, a high degree of continuity of service is sometimes demanded. To meet these needs, special 'no-break' schemes, e.g. a quick-starting generator and interposed flywheel m.g. set, or a floating battery with an invertor, are sometimes employed. Such arrangements are inevitably complicated and expensive. It is questionable whether the overall reliability will be much better than either that derived from the street mains of an electricity authority in many localities, or of the computer itself etc. Therefore, the need for special 'no-break' apparatus should be critically examined to make sure that it is really justified.

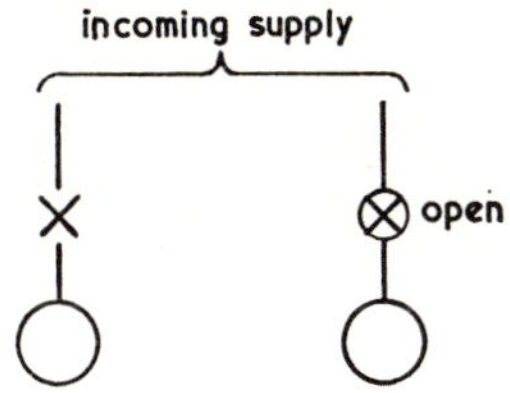

a

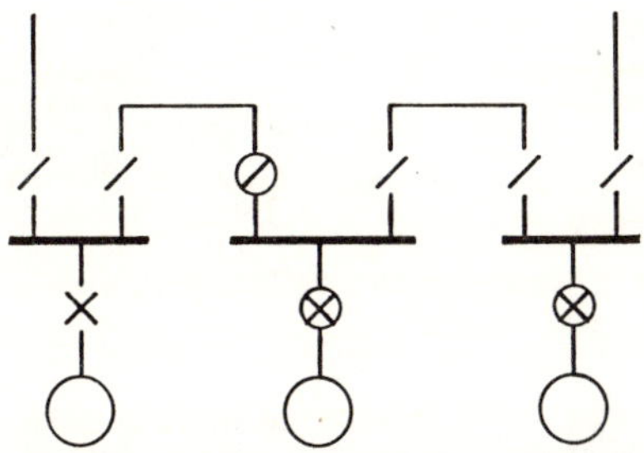

b

Fig. 15 Examples of supplies to vital pumps
a Two pumps, one running and one standby
b Three pumps, one running, one standby and one under overhaul

5.6 References

BINNS, J. W., and OUTRAM, W. J. (1959): 'Electrification of the UKAEA Industrial Group factories', *Proc. IEE*, **106**A, pp. 77–87

HUDSPETH, R. A., and MIDWORTH, N. C. (1968): 'Brush "delecta" medium voltage switchgear', *Hawker Siddeley Tech. Rev.*, **4**, (2), pp. 43–44, Figs. 4 & 5

MONEY, S. H., and HARRIS, J. (1968): 'Autoreclosing switchgear in distribution practice' *Proc. IEE*, **115**, (2), pp. 288–300

RAMSAUR, O., and TREWEEK, J. E. (1954): 'System capacity required for 3-phase arc furnaces', *Trans. Am. Inst. Elec. Eng.*, Pt. IIIB, **73**, p. 1373

6 PROVISION OF FIRM-CAPACITY HEAVY-CURRENT MEDIUM-VOLTAGE SUPPLIES

This chapter describes some of the problems involved in providing 'firm' capacity for a large concentrated block of load at a nominal voltage of 415V by using standard equipment. For this purpose, the following assumptions are made:

(*a*) The maximum short-circuit capacity of the high-voltage supply will not exceed 350MVA.

(*b*) Transformers (but not switchgear) may be subjected to an overload of 20% in emergency conditions which, it is assumed, have been determined by consideration of the ambient temperature and loading in accordance with CP1010.

(*c*) The 415V switchgear shall comply with BS3659: 1963, and fuses with BS88, Pt. 1: 1967.

(*d*) The short-circuit contribution by any motors may be neglected.

Consider first the case of two 2000kVA transformers (Fig. 16). If an open circuit voltage of 433V is assumed, the full-load current is 2680A. The largest circuit breaker available is rated at 3000A. The consequence is that the maximum load in an emergency with a transformer out of service will be restricted to 3000A. This is equivalent to 112% of the rated apparent power of one transformer, namely 2240kVA.

The transformers will need a fairly high reactance, namely 6%, to avoid exceeding the 31MVA limit, and, even so, it will obviously be dangerous to parallel two transformers. It will be noticed that there is no particular need to divide the load accurately between the two sections of busbar.

A further limitation is that the smallest rating of circuit breaker available at 31MVA is 800A. This means that the cost of switchgear will be unduly expensive for lower rated circuits if circuit breakers are specified. Therefore, for any smaller circuits, fusegear will provide a viable alternative.

The next possibility is to use three 1500kVA transformers. To keep within the limit of 31MVA, the transformer reactance will not be critical, but again the short-circuit apparent power will be excessive if two transformers are paralleled. The full-load current of a transformer is almost exactly 2000A, but rather than restrict emergency operation to this value, a 2400A breaker is chosen, which will permit an overload up to 120% with respect to the transformers (Fig. 17).

It will be possible to load each section of switchgear to 900kVA, so that, in an emergency, one of the surviving transformers will be supplying two sections; thus the total firm capacity available from the substation will be 2700kVA. However, to achieve this, accurate load sharing between the three sections of busbar is essential, and, in practice, this may be difficult to obtain (Fig. 17*c*).

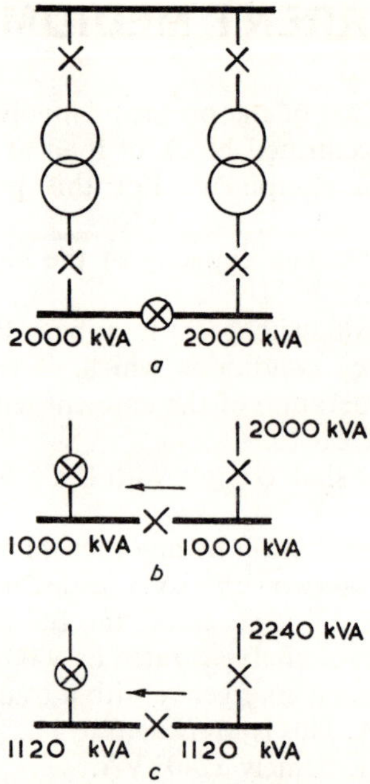

Fig. 16 Firm supplies from two 2000 kVA transformers Transformers may not be paralleled	Available capacity
a Normal	4000 kVA
b Emergency condition without transformer overloading	2000 kVA
c Emergency condition with 112% transformer load	2240 kVA

In considering the provision of firm capacity, one further possibility must be mentioned. When this is demanded at the substation busbars, it may also be necessary at each distribution board fed from the substation; i.e. each distribution board would be supplied by two feeders fed from different sections of busbar. This gives the facility of reallocating the load in an emergency, so that each transformer is overloaded to its permissible limit, which, for three 1500kVA transformers, would mean that two transformers could be overloaded to supply a total of 3600kVA (Fig. 17*d*).

The last case to be considered involves the use of four 1000kVA transformers (Fig. 18). The full-load current of each will be 1330A, and a 1600A circuit breaker would be chosen. The allowable degree of overloading would be the same as for the 1500kVA transformer, and the normal load would be 800kVA per section. The problem then arises as to whether it is permissible to parallel two 1000kVA transformers. For this rating, they are likely to be built to BEBST1 specification, and would have a reactance approaching 5%, so that, considering only the reactance of the transformers, two in parallel would exceed the 31 MVA limit.

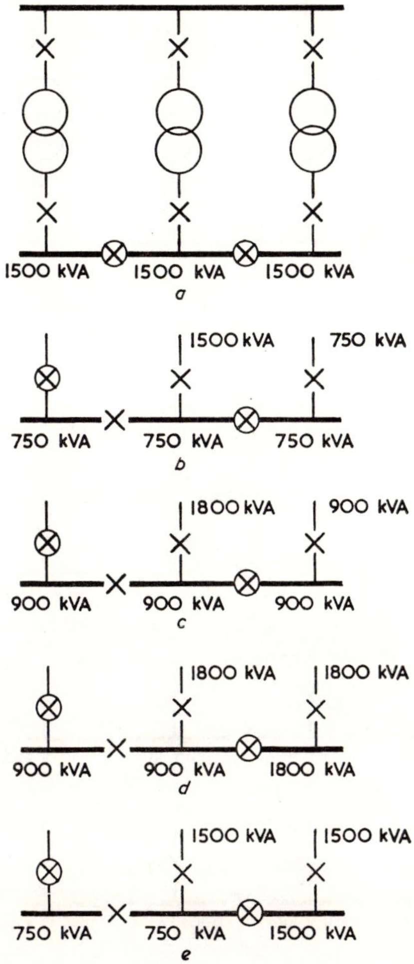

Fig. 17 Firm supplies from three 1500 kVA transformers

		Available capacity
a	Normal	4500 kVA
b	Emergency condition without transformer overloading	2250 kVA
c	Emergency condition with 120% transformer load	2700 kVA
d	Emergency condition with 120% transformer load and redistribution of load	3600 kVA
e	Emergency condition with no transformer overloading but with redistribution of load	3000 kVA

Without parallel operation, the firm capacity would be limited to 2400 kVA, whereas, if two transformers can be paralleled, the firm capacity goes up to 3200 kVA. Alternatively, if the load can be reallocated between the various

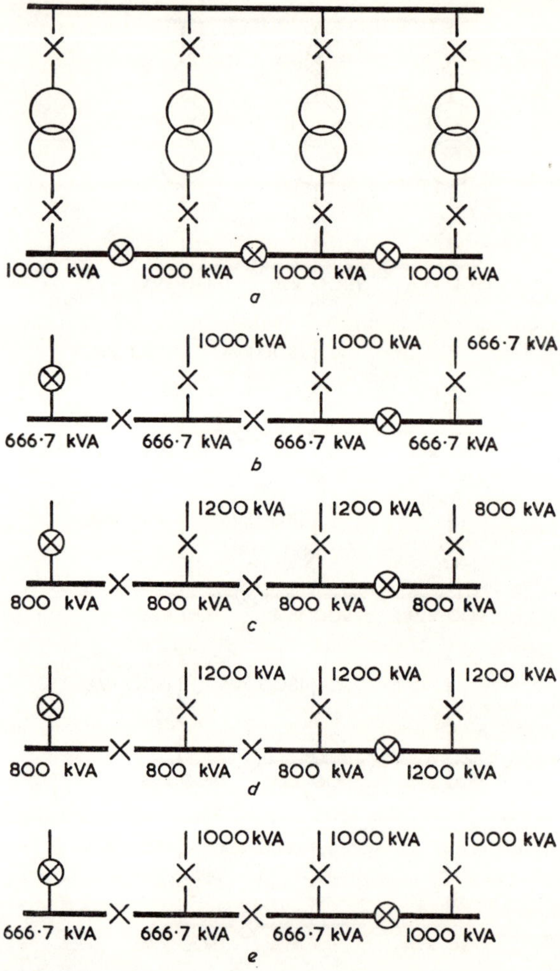

		Available capacity
Fig. 18	Firm supplies from four 1000 kVA transformers	
a	Normal	4000 kVA
b	Emergency condition without transformer overloading	2666·8 kVA
c	Emergency condition with 120% transformer load	3200 kVA
d	Emergency condition with 120% transformer load and redistribution of load	3600 kVA
e	Emergency condition with no transformer overloading but with redistribution of load	3000 kVA

sections of busbars, there is no need to parallel transformers which can each be overloaded, so that, in an emergency, a total load of 3600kVA can be supplied.

If the transformers are separated by a moderate length of busbar, and the extra reactance of this is taken into consideration together with that of the connections between the transformers and the switchgear, it is likely that the amount by which the short-circuit apparent power exceeds 31MVA will be small, and, consequently, any risks associated with paralleling two transformers may well be disregarded during an emergency. On the other hand, if it is felt that this slight risk is unacceptable, paralleling two transformers will involve no risk if all the outgoing circuits are controlled by cartridge fuses conforming to category AC80 of BS88: 1967. Under such circumstances, the fault current passing through the incoming circuit breakers and the bus sections will always be less than 31MVA.

While the cost of two 2000kVA transformers will be the lowest and the provision of four 1000kVA the highest, the latter provides some benefits, even though some are rather intangible. First, the amount of firm capacity provided is appreciably greater; secondly, under normal conditions, the fault currents will be lower and, therefore, less destructive; lastly, it is not necessary to provide all the equipment initially—it can be installed in stages to match the growth of the load.

It must be realised that the above examples are somewhat theoretical, and that, in practice, loads can seldom be subdivided to suit the available equipment. Nevertheless, the principles laid down above have a wide application, and show the limitations in the scope of application of available apparatus or multiples thereof in the alternative schemes described. Even an economy such as using a fuse instead of a circuit breaker can increase the available capacity (by permitting two transformers to be paralleled without anxiety) by as much as 50%, as was mentioned for the group of 1000kVA transformers.

It will be apparent that it is seldom possible to plan for the optimum use of each item of equipment; i.e. switchgear ratings may not exactly fit the loading which it is possible to apply to transformers. Still less is it possible to design a system for supplying a specific load without, at some point, going to the next larger size of the equipment involved. Inevitably, this will mean underrunning some part of the equipment.

It must be stressed that the examples cited here are with reference to specific parameters. It may not always be good practice to adopt such large (2MVA) transformer units, because of the large prospective fault currents. Nor may it always be considered essential to provide 'firm' capacity where relatively simple and robust equipment is involved. More smaller units without standby may often be satisfactory, on the basis that there will be 'fewer eggs in any one basket'.

7 PROTECTION

The object of electrical protection gear is to assist continuity of supply by detecting any fault in the system, so that it can be disconnected before any consequential damage arises, and to prevent any unnecessary disconnection of healthy apparatus.

Protective arrangements are available which detect almost any electrical condition that is likely to damage the equipment, but achieving perfection with them may be both costly and complicated. In practice, relatively simple equipment fulfils most needs, and more sophisticated arrangements should be reserved for applications where the extra cost can be justified.

The simplest protective device is the fuse. It is practicable to design radial systems for medium-voltage supplies by using nothing but cartridge fuses (BS88 applies) and to extend the use of such fuses to the high-voltage side of the associated transformers. In using such fuses, there are some limitations that must be mentioned.

No difficulty is experienced in obtaining enough excess current to operate a fuse when a short circuit occurs, but the same is not always true of earth faults. When fuses are used to provide earth-fault protection, an impeccable earthing system is essential to ensure that the magnitude of the prospective earth-fault current is comparable with that of a short circuit.

Direct-on-line-started cage motors will usually comprise a large part of the load to be supplied in any industrial establishment. As the starting current of an ordinary cage motor will be of the order of six times the full-load current, it will be obvious that there is a risk of fuse operation if the rating of the fuse is too small. On the other hand, the fuse must be sensitive enough to prevent damage when a heavy short circuit occurs. Vulnerable points are the cables and the heaters of the thermal overload relays. It will be obvious that, for this application, the fuse rating must be selected with care. To help the user, manufacturers publish tables giving recommendations of the appropriate fuse ratings with respect to motor powers. A typical fuse having a rating of about three times the motor full-load current may be needed, though, with some designs, fuses of lower rating may be practicable.

Although it may be, to some extent, a forlorn hope, it is the duty of the overload device in the starter to disconnect the motor if the latter is stalled or overloaded. The fuse should only operate if a heavy current due to a short circuit or earth fault should flow and which would be beyond the breaking capacity of the starter. There is a risk that an earth fault may operate only one fuse and leave the motor 'single phasing'. Consequential damage to the motor will be prevented if the overload relay is designed, as is often the case, to detect single phasing.

Where fuses are used in series, e.g. the major fuse at the main distribution board and the minor fuse for the subcircuit supplying the motor, it is important than, if a short circuit occurs on the motor, only the motor fuses operate. Where heavy fault currents are possible, as in the vicinity of a substation, it is approximately true to say that the rating of the major fuse be not less that twice that of the minor fuse. More accurately, discrimination will be achieved when the total I^2t of the minor fuse does not exceed the prearcing I^2t of the major fuse with respect to the maximum prospective fault current.

The above will be found satisfactory if the operating temperatures of both fuses are approximately the same, but, if the major fuse is fully loaded and running warm and the minor fuse is cool, a rather greater separation in the ratings of the two fuses would generally be advisable. On the other hand, at points distant from the substation where lower values of short circuits are anticipated, a smaller separation in the ratings of the major and minor fuses may be practicable. Where the characteristics of the major and minor fuses are different, e.g. if supplied by different manufacturers, their characteristics should be carefully compared to ensure that adequate discrimination is obtainable up to the maximum prospective fault current.

Fig. 19 shows how a number of fuses might be connected in series to ensure proper discrimination. It also shows how the protection and the high-voltage supply of a transformer can influence the maximum size of fuse, and hence the maximum size of cage motor that can be started direct-on-line.

One of the important principles of current interruption by cartridge fuses is that the maximum prospective fault current is always interrupted before the peak of the first halfcycle is reached, so that, in fact, the fuse can be said to have a current-limiting effect.

This principle of current limitation is being extended by some circuit-breaker manufacturers who make use of the mechanical forces produced by the short circuit to separate the contacts before the peak of the first half-cycle is reached. This introduces the impedance of the arc into the circuit which the breaker is then able to interrupt before the first halfcycle reaches its maximum, and the overall characteristic is very similar to that of a cartridge fuse.

While much of the protection of industrial medium-voltage systems can be provided by cartridge fuses or, in some instances, by circuit breakers employing current-limiting principles to achieve the same effect, the use of circuit breakers to give discrimination on a time-graded basis is often necessary on high-voltage systems, and sometimes on parts of medium-voltage systems as well. Definite-minimum-inverse-time-lag relays, or other excess-current devices that give approximately the same effect, are used.

Relays having characteristics in accordance with BS142, although suitable for many distribution circuits, do not fit very well with fuse characteristics. To get over this difficulty, relays have been developed which have a characteristic that is much more inverse than that of the BS142 relay. Another solution that is applicable to circuit breakers controlling transformers is to add a very high-set (i.e. 16 times full load) instantaneous relay, which will disconnect the transformer almost immediately if a fault occurs on its high-voltage side, and thereby make discrimination easier with the preceding high-voltage circuit breaker and at the same time give latitude for setting the induction relay or other excess-current protection with respect to the protection on the outgoing medium-voltage circuits.

Where relays are employed for overcurrent protection, no extra cost is incurred in making one of the three elements suitable for earth leakage. There are several advantages in this.

First, greater sensitivity can be obtained. This will be particularly important where the neutral of an h.v. system is earthed through a current-limiting resistance. Here the earth-fault settings of the relays may need to be different from the overcurrent elements.

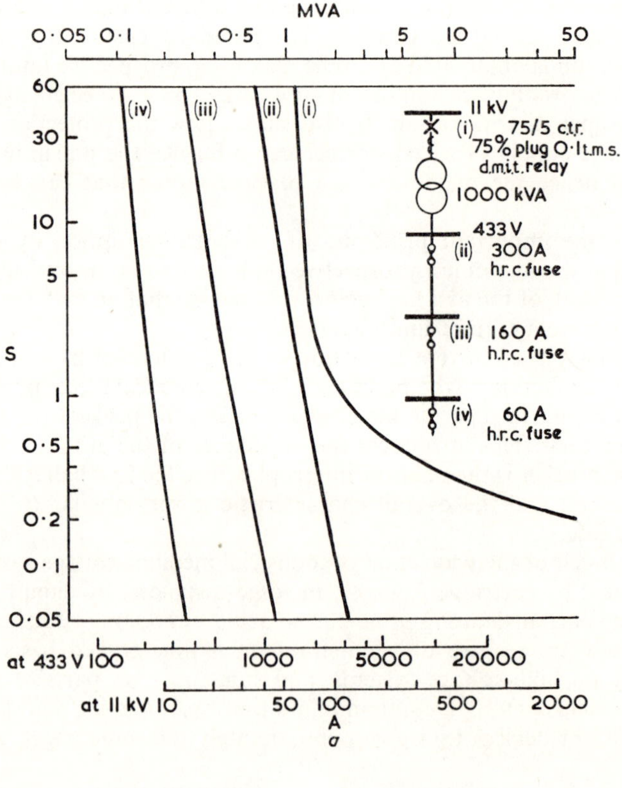

a

Secondly, it is important to remember that, as far as earth faults on h.v. systems are concerned, a stepdown transformer is a final subcircuit. This is because both phase and earth faults on the l.v. side are seen on the h.v. side as overcurrents, and will, therefore, not be detected by the h.v. earth-fault relay. Consequently, the latter can be made instantaneous provided that adequately balanced current transformers are available.

Whenever relays are provided to give discrimination on a time basis, it is worth 'drawing out' the calibration to ensure that discrimination is obtainable up to the maximum prospective values of both short circuits and phase faults. Where transformers are involved, this should be done on a common scale of megavolt-amperes, as was done for fuses in Fig. 19. With the aid of a stencil, this is not a lengthy procedure, and gives a clear picture of the result.

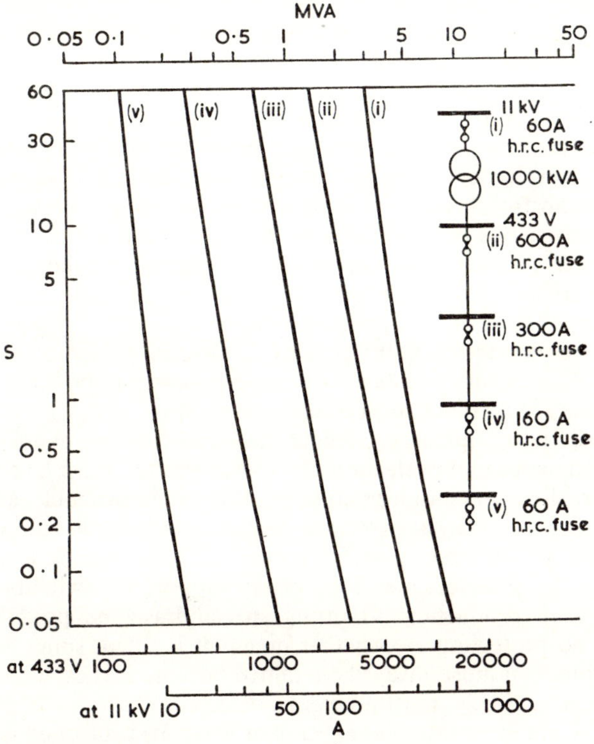

Fig. 19 Discrimination
a Between fuses and d.m.t. relay
c.t.r.=current-transformer ratio
d.m.t.=definite minimum inverse time
t.m.s.=time multiplex setting
b Between fuses only

On some systems, it may be necessary to forego some degree of discrimination, because there is not enough time available to provide adequate intervals between the steps. It will then be a matter of choosing settings which, though not perfect, will cause a minimum of dislocation if a fault occurs. Usually this will entail reducing the intervals between intermediate circuits and subcircuits, but maintaining them with respect to the supply. The object of this is that, while a subcircuit fault may interfere with other subcircuits, it does not cause a complete shutdown.

There may be exceptions to this plan, in that the importance of a subcircuit may vary with the process associated with it. The discrimination would then have to be graded so as to maintain continuity of the supply, as far as practicable, to the important load even if other supplies suffered.

Where discrimination is difficult to achieve on a time-graded basis, it is always worth checking that no unnecessary steps are included; e.g. on radial systems, it is surprising how often incoming circuit breakers or fuses to a subboard are included in the chain to the detriment of the overall discrimination.

Where adequate protection cannot be obtained on a time-graded basis, it becomes necessary to use balanced protection to give instantaneous operation. For feeders, this will involve the addition of a pilot cable. Stepdown transformers can be dealt with in the same way; alternatively, where instantaneous protection for both overcurrent and earth faults has been provided on the h.v. side (as referred to earlier), it becomes necessary to add restricted earth-fault protection to the l.v. side together with intertripping. The latter is applicable where the l.v. winding is star connected and linked to the switchgear by single-core cables, as is often the case, and short circuits between lines are a practical impossibility. Where 3-core cables are involved, overall balance must be used.

While each unit of plant, e.g. feeders and transformers, can be provided with instantaneous unit protection, the switchgear too must be considered. It may be realistic to consider that faults here are improbable, and that, if they do occur, they will cause serious dislocation of supplies anyway. Thus the addition of bus-zone protection may achieve very little extra discrimination in practice; i.e. loss of an important switchboard may effectively stop production. Omitting special bus-zone protection does not imply no protection as, whether or not it is fitted, some measure of backup protection must always be retained because it is essential that any fault, however unlikely, must always be detected.

Where the not inconsiderable expense of bus-zone protection is incurred, it is essential that whatever scheme is adopted is carefully applied. There are sometimes loopholes; for example, a fault on the orifices in conventional metalclad switchgear between the circuit breaker and the current-transformer chamber on an incoming circuit will be detected by the busbar protection, but not by the feeder protection. Thus, although the circuit breaker will trip, the fault will not be interrupted. This is a situation where intertripping between bus zone and feeder protection is needed.

Backup protection should deal with the case, fortunately rare, where a circuit breaker fails to trip even though the protection has correctly detected a fault. Not much in the way of discrimination can be expected, but it will suffice if the next step of protection in the direction of the supply detects the fault and prevents its persistence. Usually this can be achieved, but it may be difficult in some instances; e.g. if the circuit breaker controlling a small transformer (say under 500kVA) connected to a switchboard supplied from a 5MVA circuit fails to trip on the occurrence of a fault at the l.v. terminals, enough current for detection by the major circuit may not pass.

It is a wise precaution to seek the co-operation of the supply authority, since, from the largest to the smallest supply, discrimination must be achieved between the consumer's equipment and that of the supply authority. Incompatible arrangements may prove both difficult and expensive to put right later.

To ensure stability, protection on the Grid is being planned to effect clearances in very short times, e.g. 4 cycles or less, and this will naturally tend to affect the time of operation available for discrimination on the supply authorities' networks, and hence what is available for controlling the consumer's supply. As a consequence, the future needs of consumers' protection will be for faster clearance times. While this may not have much effect on the smaller consumer where current-limiting techniques, by the use either of cartridge fuses, or of circuit breakers to achieve the same effect, can clear heavy short circuits within the first halfcycle, larger users may have to face up to an increasing amount of balanced protection as an alternative to simpler time-graded systems.

As far as the distribution system is concerned, the protection of circuits to motors will need to cope with any normal starting peaks or cyclic load swings, but there are no other especial requirements. Beyond this, however, are the needs of the motor and sometimes the machinery driven by it which may call for a considerable degree of complication according to the magnitude and nature of the drive; due regard should be paid to the various statutory requirements for motor control, e.g. Regulations 11 & 12 of the Electricity (Factories Act) Special Regulations.

Fuses are extensively used on m.v. circuits and on voltages up to about 3·3kV to relieve motor contactors of the necessity of breaking heavy currents; where circuit breakers are involved, instantaneous earth fault and high-set overcurrent elements are used.

For motors, thermal overload relays incorporating a single-phase device are widely employed, and the single-phase device will detect a 'blown' fuse. Although such relays attempt to match the heating characteristics of a motor while it is running, they are too slow to follow the rapid heating of a stalled motor. For this latter purpose, the magnetic relay is better suited, though it gives little protection against overloading or single phasing. If a choice has to be made between these two types of relay, it will be necessary to decide whether stalling or overloading is more likely to occur and, in difficult cases, to provide both where the cost can be justified.

It must be remembered that motors built to BS2613 have no overload capacity, and any excess current necessary to operate a relay must cause some overheating. More sensitive protection against this can be provided by the use of thermistors in the stator winding, but such devices are impracticable in the rotor whose temperature may be the limiting feature under stalled conditions.

If the control gear is located away from the point where the motor is controlled, as happens with multimotor-control centres or with circuit breakers forming part of a switchboard, it is important that motor-protective devices (other than undervoltage) are hand reset. This avoids the risk of a plant operator repeatedly switching on to a fault condition, possibly with disastrous results.

8 EARTHING

8.1 Legal requirements

The Electricity (Factories Act) Special Regulations 1908 and 1944 apply in the widest sense to the use of electricity in factories. Of these regulations, one applying particularly to earthing may be cited:

Regulation 21

'Where necessary to prevent danger, adequate precautions shall be taken either by earthing or by other suitable means to prevent any metal other than the conductor from becoming electrically charged'.

From the definitions of the above Regulations can also be quoted:

'Earthed means connected to the general mass of earth in such a manner as will ensure at all times an immediate discharge of electrical energy without danger'.

This emphasis on 'connecting to the general mass of earth' must have led to a considerable waste over the last half century, and it is quite impossible to estimate how many earth electrodes have been provided on industrial establishments, in the form of buried ironmongery or copper rods, that serve no useful purpose.

The memorandum on the above Regulations issued by HM Factory Inspectorate (SHW928) does recognise the importance of earth-fault-loop impedance, but, having done so, goes on to stress the need for an earth electrode.

The IEE Regulations for the Electrical Equipment of Buildings attach more importance to the earth-loop impedance. Regulation D34 demands, as a first choice, a metallic return path as a means of earthing. These regulations require the consumer to fall back on the use of an earth electrode where no metallic earth-continuity conductor can provide it. The IEE Wiring Regulations, however, have little legal standing, and can only be invoked directly in the sense that, if a consumer's installation complies, the supply undertaker cannot refuse to connect the installation to the power supply (Electricity Supply Regulations 1937).

Another authoritative document, namely British Standard Code of Practice CP1013 (1965), entitled 'Earthing', also has no legal standing, and much of it is beyond the scope of industrial earthing.

8.2 Technical requirements

The reasons for earthing, which are in no way restricted to industrial establishments, are, first, to promote safety, so that there is no risk of anyone getting a shock from metalwork that is not intended to become live; and, secondly, to prevent persistent earth leakage which may cause fire.

These objects are achieved by bonding all the extraneous metalwork of electrical equipment by means of an earth-continuity conductor, so that a complete metallic circuit exists from each and every point on the system whereby any fault to earth can return to the earthed neutral point of the transformer or generator from which the supply originated.

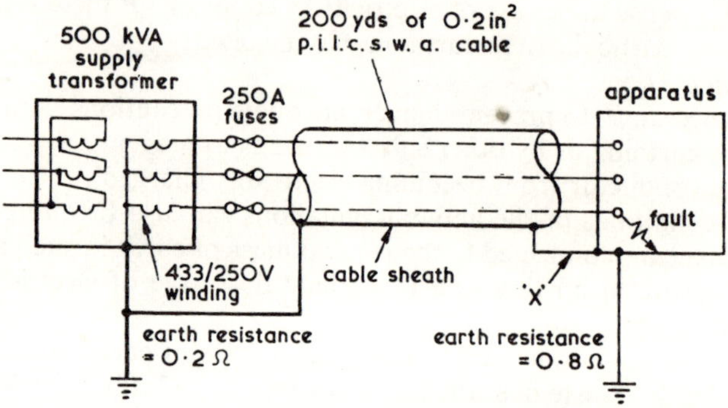

Fig. 20 Typical effect of omitting earth-continuity bonding

With bonds as shown
Fault current returns via cable sheath
maximum fault current, 2300 A
(equivalent to 0·11 Ω)
fuses blow in 0·3 s

With bond 'X' omitted
Fault current returns via earth electrodes

resistance of cable and transformer (reactance is negligible)	0·06 Ω
resistance of transformer earth electrode	0·2 Ω
resistance of apparatus earth electrode	0·8 Ω
	1·06 Ω

current=250/1·06 = 236 A
voltage rise of case above earth=189 V
even though both earth resistances are quite low
(i) fuses do not blow
(ii) fault current persists
(iii) case of apparatus remains alive

p.i. l.c. s.w.a.=paper-insulated, lead-covered, steel-wire-armoured

Fig. 20 shows the circuit traversed by the earth-leakage current; any attempt to try to make the current flow through the general mass of earth will almost certainly be doomed to failure. It will be noted that no earth-leakage current should flow through the earth electrode, and, therefore, as far as the medium-voltage system is concerned, it serves no useful purpose. As the earth-continuity conductor provides the lowest-impedance return path for leakage current, little will enter the general mass of earth. It is problematical whether 'earthing' is the best term to apply to this process, but, as no alternative exists, this well established convention will be followed.

Most industrial establishments are fed through cables where the sheath and/or armouring provides an excellent earth-continuity conductor back to the neutral of the transformer from which the supply is derived. Before proceeding further with the electrical characteristics and construction of the earth-continuity-conductor system, methods of detecting and interrupting earth-leakage currents must be considered.

The simplest and cheapest way is to rely solely on enough leakage current flowing for it to operate the excess-current protection of the circuit concerned.

This is covered by Regulation D22 of the IEE Wiring Regulations. This permits the excess-current protection to provide the necessary protection against earth leakage if the prospective leakage current is at least three times the current rating for the fuses and if the latter have a fusing factor in excess of 1·5. For fuses having a fusing factor less than 1·5 (many cartridge fuses qualify), the prospective leakage current is reduced to 2·4 times the rating of the fuse, and, for circuit breakers, the prospective leakage current must be at least 1·5 times the tripping current. In practice, little difficulty is encountered in obtaining the earth leakage impedance on industrial premises that fails to achieve three times the excess-current protection of the circuit concerned.

There is, however, one more practical difficulty in relation to the IEE Wiring Regulations. Regulation D30 demands that the earth-continuity conductor should be tested before any bonding is made to pipes or structural metalwork. This may be difficult on occasions to achieve in domestic premises, but, in factories, it is impossible. The reason is the number of fortuitous points of contacts between the earth-continuity conductor and other metalwork, apart from any deliberate bonds that may be applied later. Distribution boards, contactor panels, lighting fittings, switches etc. will be firmly bolted to the structure; so too will all categories of piping. The earth-continuity conductor will reach motors, controllers, pressure, float and limit switches on machinery and apparatus of all types that may also be served by the pipework of other services. Short of dismantling the whole of an electrical installation, which would defeat the object of the exercise, a truly independent test of the earth-continuity conductor is quite unrealistic on industrial premises. Nevertheless, this should not be considered as an excuse for omitting to measure the earth-loop impedance.

Without any attempt to provide an earth electrode especially for the purpose, the metalwork of a factory does, in fact, become just one large earth electrode at many points. The author remembers an instance, many years ago, when the principles of earth-loop impedance were less well understood, where a power company was installing a supply to a new factory. They had had some trouble at the substation in providing an earth electrode with a resistance as low as 1 Ω. Somewhat surreptitiously, they made a test to the earth bar on the consumer's installation, and were curious to know how such a low figure as something of the order of 0·1 Ω had been obtained. The answer was that, beyond impeccable bonding to ensure really effective earth-continuity conductance, there was no deliberate earthing to an electrode.

Possibly there may be small factories in the country which are fed by an overhead line without an earth-continuity conductor associated with the incoming supply, and the factory would then have to provide its earth electrode in theory. Whether a low enough earth-loop impedance for operating the excess-current protection in the event of an earth fault would be obtained is problematical. It is more than likely that there would not be enough, but, if the IEE Wiring Regulations had been followed, a potentially dangerous situation would have been detected by testing and earth-fault protection might have been needed.

There is another possibility; by the time earth-continuity-conductor system had been completed, there would be effective bonds to the steelwork and pipes, as referred to earlier, and, with these many parallel paths to the mass of earth, the area in which the factory was located would assume a sufficiently uniform potential to avoid any danger to personnel. The consumer might thus be able to sit back and not worry too much, but the supply authority, with its transformer in the corner of some nearby field, might be faced with the risk of electrocuting cattle, because the resistance of their earth electrode was much greater than that of the factory and all the voltage drop due to earth-fault current would appear at the supply authority's substation. The author must admit that he has never heard of an instance quite like this relating to industrial premises; the relations between consumers and supply authorities are usually so good that some practical way would be found for dealing with such a situation by mutual co-operation.

Two points from the IEE Wiring Regulations may be mentioned. First, Regulations D10–D13 require bonding between the consumer's earth terminal and the pipework of public gas and water services at the point of entry to the premises. Secondly, Regulation D14 demands that the exposed metalwork of electrical apparatus must be either separated from, or bonded to, any other adjacent metalwork. The spirit of these requirements needs liberal interpretation when applied to industrial premises and more especially the larger works. Here, instead of just bonding the pipework to the electrical earthing system at the point of entry to the premises, it is necessary to ensure adequate crossbonding in each building or section of the works to which the other services are brought. These requirements must also be applied in principle to all piped services, e.g. compressed air and other gases. It will probably be found that the principle of segregation is impracticable, and there will be no alternative but to bond all exposed metalwork to ensure a uniform potential throughout.

It has already been mentioned that, as far as leakage to earth is concerned, an effective earth-continuity conductor must be provided, so that the fault current returns to the neutral of the transformer from which it started; the core and tank of the transformer may also be vulnerable to earth-fault currents from the high-voltage system and an effective earth-continuity conductor must be provided for earth leakage currents flowing from the high voltage system. Some concern is often felt about bonding the earth-continuity-conductor systems associated with the high-voltage and medium-voltage systems where the tank of a transformer is common to both. Any attempts at segregation could lead to local differences of potential and possibly a shock risk to personnel. If the two earthing systems are bonded together, protection is also provided against a break-through from the high-voltage winding to the medium-voltage winding.

8.3 Earth-continuity conductors

As a general rule, the problem of providing effective metallic earth-continuity conductors is a mechanical rather than an electrical one. For this reason, cables that incorporate an earth-continuity conductor in their construction are preferred. Apart from corrosion, the only points of weakness are at the terminations and joints, so that particular care is necessary to ensure adequate bonding at switchboards and connection points.

Steel-wire-armoured cables are commonly used with either plastics- or paper-insulated lead-covered cables; in the latter, the lead sheath is bonded to the armouring at each end to supplement the continuity. Effective glands must be provided and, where there is doubt as to the earth continuity on the apparatus at the terminations, bonds must be provided. In addition, precautions must be taken to prevent corrosion of the outer sheath, and, for this purpose, p.v.c. is a convenient material.

M.I.C.S.* cables provide an alternative to the use of armoured cables and are particularly suitable for the smaller sizes. Where there is a risk of corrosion, an outer sheath of plastics material can be provided. If bending or vibration is feared, a helix can be introduced into the cable, as is sometimes done with oil pipes, which will allow this class of cable to be run to motors on slide rails.

With both armoured and m.i.c.s. cables, multiple-earthed neutral systems are possible on supplies derived from the consumer's transformer, and, in particular, the relatively good conductivity of copper sheath of m.i.c.s. cables makes this attractive for lighting circuits. Under the IEE

*Mineral-insulated, copper-sheathed

Wiring Regulations, this is restricted to single-phase supplies, but it is understood that some revision is contemplated.

Conduit systems are often found in industry, and, although many satisfactory installations undoubtedly exist, the risk of vibration loosening the conduit and/or corrosion working into the threads at the ends of the pipes often makes it unpopular, as it is felt that the earth continuity may not be properly maintained over the years. A particular difficulty is where flexibility has to be provided, because flexible metallic conduit is not in itself an effective earth-continuity conductor. Sometimes an external conductor is wound—usually rather untidily—around the outside of the flexible metallic conduit. There are two possible objections to this method; first, the earth-continuity conductor is vulnerable either to mechanical damage along its run or to a loose connection at its terminations, and, secondly, although the author has no figures available, the earth-fault current must encircle the iron of the flexible metallic conduit, which cannot but help to increase the earth-loop impedance.

If the earth-continuity conductor is run within the flexible metallic conduit, termination presents a problem; though there are some tricks, such as painting the flexible metallic conduit green to indicate that there is an earth-continuity conductor therein, there will remain the doubt that the earth continuity may not, in fact, have been provided.

Dealing with earth continuity of subcircuits may be going somewhat beyond the general theme of factory distribution, but it does help to underline the fact that earthing will not be achieved by merely connecting the metalwork of each item to the general mass of earth. A properly designed metallic earth-continuity conductor is essential.

9 SYSTEM EARTHING

Whereas Chapter 8 described the various principles involved whereby an earth-continuity-conductor system disposes of earth faults so as to prevent consequential risks due to fire and shock, separate consideration must be given to the ways in which any system is tied to earth, so that, in addition, other potentially dangerous conditions can be mitigated.

The star point of any transformer from which any supply is derived is where this is conveniently done. The earthing point of the incoming supply will be at some transformer on the supply authority's system, so that the consumer need have no further direct interest in this context (but the magnitude of the prospective earth-fault current will need consideration with respect to the protection). Where the consumer has to provide transformers, any earthing concerned with the l.v. side will be his responsibility.

The background to these requirements has two aspects. First, the supply authority will have to comply with Regulations 8 and 11 of the Electricity Supply Regulations, 1937. Secondly, the consumer must obey Regulation 20 of the Electricity (Factories Act) Special Regulations 1908 and 1944. Both these are concerned with the possible dangers of an interwinding breakdown of a transformer whereby the h.v. appears on the l.v. side.

Where consumers' generating plant is involved, rather more complicated issues are entailed, and will be dealt with later. Even where generators are concerned, it is, however, still better to use a transformer neutral, or, if this is not available, a neutral earthing compensator. The reason for this is that generators are usually taken out of service from time to time, whereas a transformer or neutral earthing compensator can be left in service almost continuously. This applies particularly to 3-phase 4-wire systems where the star point must not be disconnected from earth.

For h.v. supplies, the star point can be connected to earth through a neutral earthing resistor. This is usually designed to limit the maximum earth-fault current to about the full load of the transformer with which it is associated. The advantage of the method is that the current that can flow will cause less consequential damage at the point of fault.

A further advantage of earthing a system neutral through a resistance is that the consequential voltage disturbance with an earth fault will not be severe. Thus healthy plant, one that incorporates time lag and other loads such as high-pressure mercury-vapour lamps, will be little, if at all, affected.

E

Naturally, it is important that the protection is adequate to detect the limited amount of fault current.

For m.v. supplies, as has been mentioned, it is inadmissible to allow the star point to move away from earth potential, because single-phase supplies connected from line to neutral may be involved, and they will have switching and protection in the line, but not the neutral.

As an alternative to earthing, some h.v. systems have been successfully operated with the neutral insulated. A star/star/tertiary-delta voltage transformer is used with the star point of the primary winding earthed with a relay connected in the delta winding to detect the occurrence of an earth fault; also an overvoltage relay can be provided to detect a dangerous interwinding breakdown of the transformer providing the supply from some higher voltage. Such an arrangement would be applicable to a 3·3 kV supply fed through a transformer from an 11 kV supply, but it would be acceptable only where, in the event of an earth fault, it is possible to look for it by switching off each circuit in turn. The author has had experience of such a system where so little damage was done by a motor earth fault that, although it was difficult to find, the cause could be accurately diagnosed.

9.1 Generator neutral earthing

Where generators operate in parallel, there is a risk of circulating currents if the neutral of more than one machine is earthed at a time. This is owing to irregularities in waveform producing third and other harmonics which do not cancel out.

Possibly this risk of harmonics tends to be exaggerated, but, as their magnitude cannot be readily predicted, it is standard practice to arrange for the neutral of only one generator to be earthed. In some stations, a system of changeover switches is adopted to prevent more than one neutral being earthed simultaneously.

Where the cost can be justified, a circuit breaker in the neutral enables it to be tripped on the occurrence of a machine earth fault and prevent any further damage until field suppression has ensued. More elaborate still is a neutral busbar with a circuit breaker for each machine.

Whatever is done, there remains the risk that the machine being used to provide the system earthing may develop a fault; thus, when it is tripped, the system will lose its earthing.

Maintenance of supplies following the loss of a generator inevitably places a strain on the operators. If, as a result of such an emergency, a 3-wire system is unearthed, nothing catastrophic is likely to ensue until there has been time to take stock of the situation.

There is no objection to carrying out a test to measure any circulating current that flows in practice, though care will be needed to avoid fortuitous tripping of any earth-leakage relays that may be sensitive to third harmonics.

To determine the heating effect of any third-harmonic currents that appear at the neutral point of a machine, the total must first be divided by three, as one-third will flow into each limb of the windings; the r.m.s. of the answer must then be added to the fundamental current. The final result may well show that there will be very little additional heating. Thus, it may prove practicable to relax any precautions, provided that the effect on the protection relays referred to above are dealt with.

Where the generators operate in parallel with the public supply, that system will normally be earthed at the transformer from which the supply is derived. If the incoming supply is lost and there is again the risk that the generator will continue to operate unearthed, the considerations already discussed above will apply, except that, with a 4-wire system on medium voltage, it is necessary to provide a contactor to earth the system automatically if separation occurs.

If both generator and interconnecting transformer are adjacent, it may be possible to consider them as a unit, so that, if one trips, the other does too, but this will obviously not be applicable when the generator must be retained in service to supply some vitally important load when the supply fails.

As with transformers, a current-limiting resistance can be provided if there is any scope for restricting the prospective value of earth fault on an h.v. system, but where one of several generators is used for system earthing, this will only be practicable if the earth-fault current produced by the single machine is more than sufficient to operate the protection on the largest outgoing circuit.

If a generator is coupled to the supply through a unit transformer, a star point on the supply-voltage side of this transformer can be used for system earthing. Furthermore, if there are several such units in the station, the star point of each can be earthed. In such instances, the star point of each generator can either be earthed directly, or, alternatively, a potential transformer can be used to detect any earth leakage at the generator voltage.

10 SHORT CIRCUITS

While short circuits should seldom, if ever, occur, it must be assumed that the insulation may fail at any time and at any point on a system. It must be stressed that the risk of short circuits is not restricted to incoming and outgoing circuits; it includes the busbars and the whole of the switchgear installation. Adequate precautions have therefore to be taken for disconnecting a short circuit wherever it may occur.

Just what happens at the point of short circuit will vary with circumstances. At a 'bolted' short circuit, there will be no visible pyrotechnic display. Such faults are rare and only likely to occur when a circuit, which has been deliberately earthed as a precautionary measure, is accidentally energised.

More usually when a short circuit occurs, there is poor contact, so that arcing occurs. This will generate heat, which is damaging to equipment in the immediate vicinity and often destroys valuable evidence of the cause; in addition, the heat may cause severe burns to any personnel within range. A further effect is that a short circuit, within an enclosed space without pressure relief, may burst the chamber with explosive violence as a result of the expansion within.

The magnitude of the short circuit will be determined primarily by the constants of the circuit feeding into it; it may be restricted somewhat by the resistance of the arc, which, in turn, will be influenced by the intimacy of contact between the points of opposing polarity. For any particular short circuit, the extent of the damage will vary directly as the duration and as the square of the current.

For the purpose of selecting the short-circuit rating of the switchgear, the worst duty (from the switchgear point of view) must be assumed, i.e. a 'bolted' short circuit.

The simplified form of calculation shown in Appendix C of BS116:1952 will suffice for most purposes. For h.v. systems, the impedance of cables will usually be small enough for them to be neglected, though, if any overhead lines are involved, it will probably be necessary to include them in the calculation.

For m.v. systems, the impedance of cables and connections, including busbars, should be considered, and, where heavy-current circuits are concerned, may make an appreciable reduction in the value of prospective short circuits. In marginal cases, the rearrangement of the conductors, say the single-core cables from a transformer to the switchgear, so that they are grouped in phases rather than in trefoil, may avoid overstressing the switchgear if a fault occurs.

In calculating prospective values of short-circuit current, the effect of synchronous machinery should present no difficulty, but the contribution from induction motors is less well understood, and information on the subject has been somewhat scanty.

Cooper *et al.* (1969) give a practical approach to this problem.* It is shown that, owing to very rapid decrement of the current from the motors, the breaking capacity of the switchgear is unlikely to be seriously affected, but the 'making' capacity may have to be much greater than if the effect of induction motors were to be neglected.

This paper stresses that, on a large system, not only the high-voltage motors but also the additive effect of all the medium-voltage motors may be significant when specifying the making capacity of the high-voltage switchgear. It is suggested that, to simplify the calculations, the medium-voltage motors and the interposing transformers can be 'lumped'.

Strictly speaking, the effect of induction motors should be included in the calculation of medium-voltage short circuits. In practice, however, this is seldom, if ever, done, and, as far as the author is aware, there are no records of any mishaps attributable to this omission. However, the effect of stray impedance due to cables and connections that will be directly in series with a short circuit at any point on a medium-voltage system will tend to cancel out the effect of the motors. In addition, 'bolted' short circuits are most unlikely on medium-voltage faults, so that arc resistance will have a mitigating effect. Finally, closing on to a medium-voltage short circuit can be anticipated by a simple 'megger' test. With this background, there may not be a need to worry unduly about the motor contribution; it would be wise, however, to take cognisance of it when large (e.g. over 1 MVA) transformers and/or generators provide the supply to a medium-voltage system comprising mainly motors for which the switchgear is near the limit.

*COOPER, C. B., MACLEAN, D. M., and WILLIAMS, K. G. (1969): 'The application of test results to the calculation of short-circuit levels in large industrial systems with concentrated induction-motor loads', *Proc. IEE*, **116**, (11), pp. 1900–1906

11 GENERATOR PROBLEMS

The methods of operation of generators for supplying industrial distribution systems fall into several categories, namely

(*a*) to provide the total supply

(i) where it is economical to do so, usually because either suitable waste heat is available or the generation cycle can provide heat for process

(ii) because of geographical remoteness of public supply mains

(*b*) to provide part of the total supply up to the economical limit of *a*(i) above

(*c*) to provide an intermittent supply to deal with emergencies and/or peak lopping.

Several operational problems must be mentioned.

In an isolated station, a decision will be necessary as to whether a spinning standby is justified. If this cannot be done, immediate load shedding will be necessary to maintain supplies following the loss of a generating unit. Here, it will be obvious that the least important circuits must be disconnected. This will not be easy to do if the outgoing circuits each contain a mixture of vital and unimportant load. Thus some segregation of loads with respect to their order of importance will be necessary.

This problem is not restricted to isolated generating stations, but will apply where a part only of the load can be supplied from the available generating plant.

There are, however, some further complications where a consumer's generating plant is operated in parallel with an infinitely larger public supply network. The power output will depend on the energy input to the prime mover, but the speed will be determined by the frequency of the public supply. The energy input must therefore be regulated in some other way: e.g. where a condensing turbine draws steam from a waste-heat boiler, it must consume (up to its maximum) all the steam that is produced; alternatively, with a back-pressure set exhausting to a process, the needs of the process determine how much steam is passed through the set.

In both examples, an immediate change takes place if the public supply is lost, and the generator is left running on its own. The control of frequency is lost and the generator governor must take over and maintain the frequency within reasonable limits. At the same time, as has already been stated, load will have to be shed to restrict it to the limit of the available generation.

In the first instance, there may not be enough waste heat available to provide enough steam for the load that it is wished to maintain. The deficiency must be made good by the provision of extra heat, e.g. emergency oil firing. On the other hand, any surplus heat can be released directly to atmosphere during the emergency.

A back-pressure set can continue to operate satisfactorily when the process steam demand is in excess of that needed to generate the required power, provided that the excess can be bypassed through a reducing valve. Where the process cannot take all the steam passing through the turbine to generate the required power, the surplus must be blown to atmosphere (which is not very satisfactory), or there must be some other outlet for the excess steam, e.g. a dump condenser.

Rather similar problems apply with respect to the control of excitation. Operating in parallel, there is only a little scope for a voltage regulator to effect any control. Obviously, whatever is done on the consumer's system will not alter the voltage of the infinitely large public supply network, although the voltage of the latter may vary from time to time. Varying the excitation of the consumer's set will only affect the voltage by varying the drop through the impedance (usually the reactance of transformers) linking the consumer to the public supply. This can only be done by altering the power factor where there is little scope for variation without exceeding both commercial and technical requirements. Hence it becomes necessary to use other methods.

Fortunately, the static regulator possesses considerable flexibility. It is possible, for instance, to arrange for a generator to operate at either constant power factor or reactive power while in parallel. If separation from the public supply causes voltage fluctuation, a high/low-voltage relay operates and switches the regulator to control the voltage.

The older type of 'mechanical' regulator can often be used successfully, so that, under normal parallel conditions, the adjusting rheostats are set so that the generator is fully excited and the regulator is just against the stop trying to raise the voltage; thus it has a setting of a few per cent. above normal. If separation occurs, the machine will then regulate the voltage to this slightly higher value. This method is applicable to fairly steady loads and where a maximum amount of power-factor correction is required.

The general principles relating to generator protection do not differ greatly from those applicable to large central stations, but many details require somewhat different treatment.

First, the units concerned are generally smaller, so that some degree of simplification may be needed to keep the cost within reasonable limits. For instance, at medium voltage, restricted earth-fault protection may suffice instead of overall current balance.

The switchgear in a small station is unlikely to be provided with bus-zone protection, so that it will be necessary to make do with backup-overcurrent protection provided by definite-minimum-inverse-time (d.m.i.t.) relays. Owing to the severe decrement with a nearby fault, the operation of such relays is, at best, somewhat problematical when it is sought to ensure adequate discrimination with outgoing circuits. This problem is made even more difficult when, as is often the case, it is necessary to consider both the maximum and the minimum amount of generating plant that is run to suit variations in the station load.

Some help in this direction can be achieved if voltage-biased overcurrent relays are used. A severe drop in voltage, such as would occur with a nearby fault, releases a low-voltage element in the relay that increases its sensitivity with respect to overcurrent and enables it to operate satisfactorily despite decrement.

If an electrical fault in an alternator is detected, it is desirable to suppress the field, but this is impracticable where brushless excitation is employed; furthermore, with an earth fault, the neutral should be broken if the machine is being used for system earthing.

One other form of protection, which can be applied to a small station, is that the prime mover should be tripped if a fault occurs so as to reduce the dependence on the overspeed protection afforded by the governor.

Where, as has been mentioned, generators have to operate in parallel with the public supply, the consumer's generation must be capable of maintaining the supply to any vital loads and as much other important load as is possible without overloading when there is an interruption either of the public supply or at some intervening point on the consumer's system. In this contingency, automatic separation is essential, and this must be arranged at some point on the consumer's system where the load from the generator and the public supply approximately balance or, if this cannot be achieved, there is a small 'export' from the generator.

If such a failure is preceded by a short circuit, it can readily be detected by directional overcurrent protection with fast and sensitive settings for the flow of current in the direction of the public supply. Fortuitous tripping in the event of a short circuit that does not involve a total shut down of the outside supply does not matter; the generator continues to supply its share of the load, and can be resynchronised.

The nature of the trouble may be such that it cannot readily be detected by directional overcurrent protection; e.g. there may be an earth fault or accidental tripping at some other point on the network. To deal with this, it is possible to employ underfrequency protection.

To sum up, it is important to remember that a state of equilibrium must always be maintained for both power and reactive power. If the source of energy, i.e. the supply authority or the consumer's generating station, or a combination of both, cannot provide enough power to balance what the load demands, the result will be manifest by falling frequency; and, of course, vice versa. The same is true of reactive power, where a lack of balance will become apparent by disturbance of the voltage. It is only under exceptional circumstances that the supply authority may have any difficulty in meeting the consumer's requirements, but it will be obvious that, where the latter is not available, the capabilities of the installed generating plant must not be exceeded.

12 EFFECT OF POWER FACTOR

It is assumed that the reader is conversant with the vectorial calculations relating to power-factor correction and the effect of low (lagging) power factor on voltage regulation.

Nearly all tariffs for the supply of electricity include some form of inducement to correct low power factor or alternatively stipulate a minimum, e.g. 0·85 lagging. Commercial considerations will dictate how far it is necessary to take steps to provide correction, but it will be obvious that, where there is a high apparent-power demand charge, it will pay to provide improvement almost to unity.

In addition, a low power factor can affect the design of distribution systems by demanding an unnecessarily large installed transformer capacity and extra feeders not only to carry the reactive (wattless) component of the current but also to counter the voltage drop. Also, where power is generated, the apparent-power ratings of the alternators may have to be increased.

The ideal to be aimed at is that low power factor should be corrected at source, though there are various economic and practical difficulties that may prevent this either in whole or in part.

For large (e.g. about 3000 hp and over), continuously running drives, the synchronous motor is often cheaper and more efficient than the equivalent induction motor, because the lower apparent-power rating of the stator permits a smaller machine. Below this, the induction motor becomes cheaper in first cost, but the cost of any necessary power-factor correction by capacitors must be added; this, coupled with the slightly higher efficiency of the synchronous motor, may make the latter viable down to about 200 hp.

A further practical point is that induction motors that are too big for ball and roller bearings must have a larger airgap, and will therefore operate at a lower power factor; hence the speed and frame size of the motor may become relevant.

While unity-power-factor synchronous motors may provide the best solution for a particular drive, it is often necessary to look at the problem in greater depth. It is possible to build synchronous motors for any leading power factor, and, in the extreme, synchronous machines can be used entirely for power-factor correction with no attached load. The latter are often referred to as synchronous compensators.

The losses of a machine operating at leading power factor will be greater than at unity power factor, and the machine itself will cost more. It then becomes necessary to compare the cost of the extra correction, obtained by operating at leading power factor to offset other lagging load, with the similar degree of correction that could be obtained by the use of capacitors.

Where generators are involved, the machines can be designed, though again at extra cost, to operate at any lagging power factor, and it then becomes necessary to determine whether the extra cost of the generator would have been better spent in correcting power factor of the load by other means such as synchronous motors or capacitors.

If synchronous machines are used, their contribution to prospective short circuits must not be forgotten, and installations comprising a high percentage of synchronous motors may pose problems with the breaking capacity of the switchgear.

The starting current imposed by a synchronous motor may cause difficulties with the voltage regulation of a supply system, particularly if there are any loads which are especially sensitive to voltage. Provided that there are no special loads, such as computers, to be considered, some voltage drop is admissible, and it may be simpler to make special arrangements for the computer, as the problem of voltage variation due to starting motors of all types will have to be dealt with anyway. Drives powered by synchronous motors will be started relatively infrequently, so that any 'dipping' of the lights, though noticeable, will not cause undue irritation. Reference to the maximum permissible voltage variation is made in Chapter 5.

For drives that can run up to speed against a torque that is low compared with that needed at full load, e.g. motor-generator sets, series reactor or Korndorffer autotransformer starting can be employed if direct-on-line starting embarrasses the distribution system. For other drives that are difficult to start, because of either the high starting torque required or the high inertia of the load, synchronous induction motors provide practical alternatives to the ordinary synchronous motor.

Although there are many applications for which synchronous motors can usefully be employed, it is probably true to say that most power-factor correction is provided by means of static capacitors. For continuously running plant, unit capacitors connected to the line terminals of induction motors provide a simple and economical arrangement, as no additional control gear is required for the capacitor.

There are some limitations to the use of unit capacitors with continuously running drives. If too large a capacitor is connected to the terminals of a motor, self excitation, possibly causing overvoltage, will occur, if the supply is disconnected. As a general rule, British Standard Code of Practice CP321.102 (1950) should be followed. This restricts the rating current of the capacitor to 85% of the normal no-load magnetising current of the motor.

The no-load magnetising current is related to the motor design, and will depend on the airgap, the speed and the arrangement of the rotor. Capacitor manufacturers publish tables of maximum permissible capacitor sizes in relation to typical motor designs. These tables are sufficiently accurate for practical purposes, so that it is seldom necessary to refer to the motor manufacturer.

When an induction motor is disconnected from the line, the voltage at the motor terminals will decay approximately exponentially, and take up to about 1 s to approach zero. After about 0·25 s, the voltage may be as high as 50%. If the motor is reconnected to the line while it is still generating—as may happen in a break-before-make type of changeover starter, e.g. star–delta or change-pole—it is fortuitous whether reconnection takes place in or out of phase. If the latter, there will be a high transient current, which may damage the motor.

If the motor is fitted with a unit capacitor, self excitation is prolonged for perhaps several seconds, and the chances of producing a high transient are correspondingly increased. The magnitude of the transient will be even greater if the motor is effectively running above synchronous speed as with an overhauling load or when lowering speed on a change-pole motor.

For the above reasons, the use of unit capacitors should be avoided on break-before-make starters unless approval is sought from the motor manufacturer with respect to both the rating of the capacitor and its method of connection.

For motors of 30 hp and under, trouble is unlikely with simple star–delta starters if six terminal capacitors are used, so that one capacitor element can be solidly connected to each limb of the motor winding. However, for motors of such a low horsepower, it is usual to employ direct-on-line starting and avoid the complication and cost of the star–delta method.

There is no upper limit to the size of the motor for using unit capacitors. They can be added to existing induction motors (both cage and wound rotor) including h.v. machines where synchronous motors would otherwise be applicable. In the downward direction, it may be convenient to fit any small continuously running motors with unit capacitors if they can be accommodated, but approximately 25 hp is a convenient lower limit for unit capacitors.

When unit capacitors are fitted, the full-load current as seen by the overload releases in the starter will be lower, and these should be adjusted accordingly.

It will be appreciated that power-factor-correction equipment fulfils a useful function only so long as plant with which it is associated operates. The problem sometimes arises where standby equipment is installed. There may be two compressors of which only one runs. If both were to be provided with synchronous motors, the extra cost of one of the synchronous machines would always be wasted. Here there may be a case for using induction motors instead and providing one set of capacitors separately switched for correcting the power factor of the running compressor.

A capacitor fitted to a motor that is only used intermittently may serve little useful purpose, but a common capacitor for a group of such motors where considerable diversity applies will give much better value for money.

Apart from motors, unit capacitors can usefully be employed for any plant that operates continuously for long periods, such as induction heaters. It is also standard practice for discharge-lighting fittings to incorporate capacitors that will raise the power factor from about 0·5 to 0·85. A warning is necessary here, as any attempt to provide oversize capacitors on discharge-lighting circuits to correct other lagging power factor will provoke excessive neutral third-harmonic currents.

The first line of defence against low power factor should always be to carry out the measures outlined above where applicable. Often, more will be needed, and recourse must be made to bulk power-factor correction. Several problems are involved.

Where possible, bulk power-factor correction should be done as near to the source as possible so that the apparent power to be transmitted from the point of supply to the point of usage is a minimum. On the other hand, the automatic equipment necessary for the control of power-factor-correction capacitors is expensive and, as far as the commercial purchase of electricity is concerned, can be conveniently located and controlled adjacent to the point of supply.

To deal with these conflicting requirements, some compromise is necessary. First, some nonautomatic power-factor correction can conveniently be provided at each main distribution board which supplies load at low power factor and which cannot be corrected by other means. By nonautomatic is meant capacitors that are left continuously energised. If this is done, some voltage rise must be expected at times of very low load, and a practical limit is that the total reactive power of capacitors so provided should not exceed 10% of the installed transformer apparent power for the supplies derived from any substation. Any leading power factor caused by continuously energised capacitors, as described above, will be innocuous, and, in any case, is likely to result in no more leading reactive power than an automatic-power-factor-control relay would before cutting out one stage of correction.

Lastly, if more power-factor correction is still required to achieve the desired result, automatic power-factor control can be provided as near as possible to the point of supply where a control signal corresponding to the payment metering can conveniently be obtained. The automatic control equipment would comprise a multielement control relay impulsing to contactors that cut in or out capacitors in convenient steps so that the power factor is maintained at the desired value, which, for apparent-power tariffs, will be at or near unity.

As an alternative to the above in a large works with a number of substations, separate automatic-power-factor-control equipment can be provided at each substation. This arrangement will be more expensive because, in aggregate, more power-factor-correction capacitors will be needed, as the correction may have to be provided at different substations at different times, and, furthermore, additional control relays will be needed. Such an arrangement will, however, have the advantage that the load on individual transformers is kept to a minimum, and may effect appreciable saving in copper losses in heavily loaded units.

Switchgear used for the control of capacitors needs careful consideration, and most manufacturers apply appreciable derating factors. This is necessary because the current to a capacitor may be flowing without intermission for long periods, and, in addition, overloading due to harmonics and/or overvoltage at time of low load is possible. In addition, breaking the leading current to a capacitor is not an easy duty for any circuit breaker, switch or contactor.

Some systems supply plant that generates a considerable number of harmonics. These appear at the terminals of the capacitors as distorted waveforms, which can be resolved into the fundamental and a number of higher harmonics. As the reactance of a capacitor varies inversely as the frequency, the current may be excessive if an undue proportion of harmonics is present, and the consequence will be damage due to overheating. BS1650 permits a capacitor to exceed its rated current owing to both overvoltage and harmonics by 15% in aggregate.

Where the current limit is likely to be exceeded by the presence of harmonics, it is sometimes practical to feed the capacitor through a choke that has only a low reactance at supply frequency but a sufficiently high reactance at the harmonic frequencies to restrict their entry into the capacitor to an acceptable level. With such an arrangement, the resonant frequency of the choke and the capacitor is kept away from any harmonic frequency.

As a transformer will have the same effect as a choke, it is often practical, to operate capacitors satisfactorily, if they are connected to a part of a system that is separated from the source of harmonics by a transformer.

Some systems such as those supplying silicon-controlled rectifiers may contain a high percentage of harmonics, and capacitors for power-factor correction may present a difficult problem. Careful study is essential to analyse the likely magnitude of the harmonics and to devise a solution.

Networks of rejector and acceptor circuits are likely to be needed. These must be arranged so that they prevent the harmonic currents from entering where they may cause damage but allow them to flow where they can be dissipated harmlessly.

Such arrangements are inevitably costly, so that, where high power factor is commercially important, there may be a case for synchronous condensers instead of capacitors. An alternative may even be to substitute conventional Ward Leonard sets with leading-power-factor motors for some of the drives that would otherwise be supplied by silicon-controlled rectifiers. What then seems important is that such an installation must be considered as a whole, and that it is not sufficient to consider the variable-speed drive with its silicon-controlled rectifiers as a separate entity and then try to devise some means of correcting the power factor as an afterthought.

13 CABLE PROBLEMS

For cables, the choice of the metal for the conductors will lie between copper and aluminium. Because of widely varying prices, comparisons must be made at frequent intervals. Whereas copper is easier to joint and less liable to set up electrolytic action under damp conditions at the terminations, techniques applicable to aluminium are now satisfactory and, properly applied, should give no cause for anxiety. For aluminium, solid (as opposed to stranded) conductors are available for voltages up to 3·3kV.

Beyond the conductor, there is, at first sight, rather a bewildering selection of materials available for insulation, sheathing and armouring. Comment will therefore be restricted to a few of the more popular types.

For both h.v. and m.v. circuits, multicore paper-insulated lead-covered cable with steelwire armouring and an overall p.v.c. sheath has proved satisfactory for most purposes.

Various plastics may be used in place of paper, but the only widespread use has been of p.v.c. for m.v. applications.

For paper-insulated cables, a continuous sheath of lead is normal practice. The use of aluminium is an alternative; it is stronger than lead, and would probably be used more frequently without further protection but for corrosion difficulties.

Multicore lead-covered cables are usually provided with steelwire armouring. Double steeltape armouring provides slightly better mechanical protection, but has a higher electrical resistance, which makes it less satisfactory as an earth-continuity conductor. An overall sheath of p.v.c. provides adequate protection against corrosion under most conditions, and is reasonably resistant to fire. Possibly the odd puncture may occur in service, but not enough to permit serious deterioration of the armour.

For p.v.c.-insulated cables, no metallic sheath is needed, as the insulation is inherently resistant to moisture. Armouring is still necessary to provide both earth continuity and mechanical protection.

For heavy-current connections, such as on the l.v. side of transformers in substations, single-core cables will be needed. Although paper-insulated lead-covered cables with overall p.v.c. sheath are satisfactory, viable alternatives are to employ butyl-rubber insulation with a neoprene sheath or mineral-insulated copper-sheathed cables.

Obtaining suitable routes for main distribution cables is often a difficult task. Unless the conditions in the soil are exceptionally bad, buried cables are reasonably satisfactory. In addition to the need to keep careful records, markers at strategic points along the route are helpful. Also cover slabs a few inches above each cable should be provided as an additional protection. Even so, a buried cable route remains somewhat at the mercy of mechanical diggers. One limitation of a buried cable route is that severe derating is necessary if more than two or three well spaced cables run together; more than one layer of cables is in many cases unsatisfactory.

For a number of cables, one method is to provide a culvert. The cables can be run on hangers on each side and some on the bottom. The culvert can be completed so that it is flush with the ground by using removable reinforced concrete slabs. The disadvantages of such an arrangement are that the culvert tends to fill up with dirt and rubbish, stagnant water may accumulate, and vermin may be troublesome. Some ventilation is needed and, under a tropical sun, there will be a tendency for such a culvert to become an oven. Nevertheless, given a certain amount of attention, this can be a satisfactory method.

Running cables on gantries is expensive where one has to be provided especially for the purpose, but sometimes cables can be conveniently run on gantries provided for other purposes. Racking at ground level is sometimes used in chemical works to avoid burying cables in bad ground, but the routes follow pipe reservations, so that the cables are not in jeopardy from extraneous hazards, such as dumped material, vehicles etc.

Where a large number of cables have to follow a particular route, a cable tunnel may be the only, if expensive, solution. Two points require special consideration; first, the heat liberated by the cables may call for forced ventilation. Secondly, although perhaps a remote risk, some consideration must be given to fire; water sprays are probably best, so that any trapped personnel escape with a ducking at the worst. A cable tunnel could be shared with some other services but not steam or gas pipes. Convenient access at each end and adequate lighting are obvious necessities. Detection of fire in a cable tunnel is by no means straightforward, and requires careful consideration.*

A consideration with respect to paper-insulated cables is whether nondraining cables should be used to avoid messy conditions, which prevail when there is a difference in level and the cable insulating oil softens the compound in the cable box at the lower end and overflows.

Apart from the mess, there is no objection to this, but it can be avoided by standardising on mass-impregnated nondraining (m.i.n.d.) cables, the impregnation of which has a melting point higher than the working temperature of the cable. Should there be an imperfection in the lead sheath, with this type of construction it is likely to pass unnoticed, but the risk of this happening must be considered small. Another possibility† is that entrained air may be expelled at the ends, causing voids at

*O'DOGHERTY, M. J., YOUNG, R. A., and LANGE, A. (1971): 'Detection of fires in cable tunnels', *Elect. Rev.*, 26th Feb., pp. 263–266

†WITTS, A. C. G. (1966): 'Electrical distribution in papermills' (Pergamon), p. 39

the terminations. In practice, negligible troubles are encountered with m.i.n.d. cables; the only case for restricting their use is in situations where there are differences of level at voltages of, say, 11 kV and over.

The publicity literature of the leading cable makers gives adequate information for selecting the proper size of cable for any particular load, with respect to the method of running.

Derating factors relating to grouping and abnormal ambient temperatures must be carefully obeyed.

In addition to selecting a cable that will carry the required load, the magnitude and duration of the maximum prospective short circuit must be considered to ensure that the cable does not suffer any consequential damage.

It may prove in practice that most severe derating may apply over only a short length of the route, e.g. the point of entry to a substation. Here, careful dispersal of cables may allow a reduction to be made in the size of several circuits. An alternative, for long feeders of more than one drum length, is to increase the size of the length at the substation end. When considering derating due to grouping, some cognisance may be taken of the overall loading of a group. For instance, if two cables leaving a main substation run together for some distance before parting company to form a ring main, no derating would be needed, as only one cable would be required to carry the full current in an emergency; normally the current would be shared between them.

14 INSTRUMENTATION

14.1 General

There is really no limit to the number of instruments that can be fitted to a switchgear installation. Therefore, some compromise is necessary between excessive first cost and complete lack of operating information that would ensue from omitting all instrumentation.

It is useless to provide instruments that are beyond the comprehension of those who have to assimilate the data provided. Also, an unnecessarily large array of dials may be bewildering, and have the effect of masking that which is really important.

Despite these rather conflicting requirements, it is possible to lay down a number of useful general guidelines.

14.2 Ammeters

The use of ammeters should be restricted to important circuits. Just what is to be considered important may be hard to determine. In practice, this will usually mean the larger circuits, i.e. all at high voltage and over, say, 300–400 A at medium voltage. Quite small circuits may sometimes justify an ammeter when their importance is out of all proportion to their magnitude.

In the interests of economy, an ammeter may have to share current transformers with protective devices; if this is done, selector switches should not be used.

Appreciably out-of-balance currents may be found on medium-voltage 3-phase 4-wire circuits where single-phase loads are involved. A convenient place to check the overall balance is on the medium-voltage side of each transformer involved. For this application, three ammeters, compared with one meter and a selector switch, will give a better picture of a varying load (e.g. welders). If space on the panel is a difficulty, miniature ammeters may provide a solution, and the overall cost will be little more than that of using a full-sized instrument and a selector switch; in addition, the way will be left clear for using the current transformers for restricted earth fault or other protection.

Many, if not most, substations are left unattended, so that any high sustained loads might well pass unnoticed if only ordinary indicating ammeters were to be provided. Where such conditions are feared, there will be a case for providing thermal delay ammeters with 'maximum' pointers as an alternative, or in addition, to indicating ammeters. For transformers, a simple compromise would be a thermal delay ammeter on the h.v. side and ordinary ammeters on the l.v. side.

Care is needed in choosing suitable scales for ammeters. The maximum deflection should be somewhat in excess of the permissible load of the circuit. For instance, a 1000 kVA transformer having a full load of 1330 A at 433 V might be connected to the system through a switch rated at 1600 A. If an ammeter graduated 0–2000 were to be provided, the top 400 A of the scale would be serving little useful purpose. In practice, this may not matter if the upper part of the scale is partially suppressed so that the top of the open part of the scale is at about 1400 A and thereby the ammeter is easy to read for the normal loads of the circuit. A red mark at the nominal full load (1330 A) can be helpful, provided that it is realised that the transformer referred to may well be capable of carrying overloads up to the full rating of the circuit breaker.

14.3 Voltmeters

For high-voltage circuits, the cost of the necessary potential transformer will be many times that of the voltmeter itself. Voltmeters should therefore be avoided except where they are absolutely necessary, unless a potential transformer has had to be provided for some other purpose.

Where the knowledge that a circuit is alive is all that is needed, a cheap alternative is the provision of a set of neon indicators operated from capacitors incorporated in bushings.

For transformers providing medium-voltage supplies, a more convenient, and often more useful, place for the voltmeter is on the low-voltage side of the transformer, thereby obviating the need for an h.v. potential transformer and, at the same time, giving a better indication of the voltage with respect to the utilisation of the supply.

Some care is needed in selecting the point (or points) on a switchboard at which the voltage is sensed. While a single indication of busbar voltage may suffice, a better alternative is to provide indication on the circuit side of each incomer. This is helpful when carrying out switching operations, and, in the extreme, may prevent mistakes on complicated systems. In addition, unnecessary appendages to the busbars are avoided, and, where 3-phase potential transformers are involved, a ready means for phasing out is provided.

14.4 Kilowatt meters

Kilowatt meters have few applications in industrial distribution systems, as ammeters usually give ample indication of power flow. Where it is important to know the direction of power flow, a power-factor indicator in conjunction with an ammeter will give the same information, and, at low values of power factor, a better indication of the loading of the circuit concerned.

In practice, kilowatt meters are mainly used for consumers' generating plant to indicate the load on the prime mover.

14.5 Synchroscopes and frequency meters

Such instruments are not required except where some, or all, of the consumers' power requirements are provided by their own generating plant.

Synchroscopes are unnecessary where small synchronous generators have to be paralleled with the public supply. A simpler procedure is to run the set up to speed, close the stator switch and then energise the field; after this, the machine will pull into step in the same way as a synchronous motor does on starting.

In the UK, frequency variations of the public supply seldom occur, so that a frequency meter is of little use except where plant has to be run out of parallel. Prior to, and after, synchronising, a tachometer, if conveniently located, will give a good indication of speed, and, where a synchroscope is available, it will show whether the speed is too high or too low. In some systems, considerable drift of frequency may occur, and a frequency meter then has obvious advantages. If it appears likely that variations in frequency may be responsible for upsetting the operation of the plant, a recording, rather than an indicating, frequency meter may provide better information when sifting the evidence after a disturbance.

14.6 Kilovar meters and power-factor indicators

When studies relating to load flow have to be undertaken, it is often necessary to consider power and reactive power separately. At first sight, a kilovar meter might seem the first choice for strategic points on an interconnected system. However, any important circuit will incorporate, first, an ammeter to indicate the load (or overload) with respect to the heating of the circuit concerned; the addition of only a power-factor indicator will enable all other information to be deduced, i.e. the direction of power flow will be shown by which sector the instrument is reading; the power and reactive power can then be calculated.

In common with a voltmeter, a power-factor indicator needs a potential transformer if it is to be used on a high-voltage supply. Here again, it may often be economical to relegate power-factor indicators to the low-voltage side of each transformer.

For synchronous rotating plant, either a kilovar meter or a power-factor indicator will be equally convenient for adjusting the field to show the effect of the machine on the system power factor. For this application, there is little to choose between the two types of meter, but a necessary adjunct is a field ammeter, which, with rotating exciters, should be of the centre-zero type.

14.7 Kilowatt-hour meters

A kilowatt-hour meter serves two main purposes. First, the consumer may provide his own meter as a check on the supply authority's metering. Metering equipment is very reliable, and, if the supply authority's metering should fail, it is usually possible to estimate the consumption without acrimony. Nevertheless, for big supplies, where considerable sums of money are involved, the record of a consumer's checkmeter, even if it does not show the demand, will provide useful evidence for agreeing the consumption.

The second, and often important, use for kilowatt-hour meters is departmental costing. A kilowatt-hour meter is yet another item requiring a potential transformer if it is to be used on high-voltage circuits. As, for this particular purpose, the accurate assessment of transformer losses is seldom of significance, a medium-voltage kilowatt-hour meter will be less costly to install than a high-voltage one. Again economy can often be achieved by the use of balanced load metering and sharing the use of protective-current transformers.

By simple metering on the above lines, it is practicable to allocate the cost of electricity between the departments on the basis of the average cost of each kilowatt-hour purchased. For most applications, this is adequate, but, if it should be considered inequitable, each department's metering can be a replica of the supply authority's equipment, and each department can be charged on the basis of the supply authority's tariff. This may involve considerable expense in metering equipment, and a possible result is that, owing to a diversity of the demands of the individual departments, the aggregate of the departmental costs will exceed the overall cost of the purchased electricity supply, and the factory electricity distribution will show a fictitious profit.

15 SAFETY

The dangers associated with high-voltage electricity need no elaboration, but safety procedure is sometimes omitted through ignorance or negligence. It is the purpose of this chapter to emphasise the principles involved and the routines that should be adopted to prevent accidents.

It is not intended to imply that precautions are unnecessary with other electrical equipment; great care is always necessary. There must, however, be some limit to which the detailed precautions described here are relevant. It would be so irksome to apply them to an electrician changing a lamp in an office that they would fall into disrepute and be neglected. Although it may not always be easy to draw the line, it is suggested that the principles in this chapter should apply to all equipment, irrespective of voltage, in any substation where high voltage is present. It may be necessary to extend the jurisdiction further where the risks justify it.

15.1 Legal requirements

On industrial premises, the Electricity (Factories Act) Special Regulations, 1908 and 1944, apply. It is a statutory requirement that these regulations should be posted in all premises, and Regulation 29 further requires instructions for the treatment of electric shock to be posted. It is not altogether clear what is meant by premises, and, whereas one copy of each should always be posted along with other statutory notices that have to be exhibited in each factory, it would be a commonsense precaution to post the instructions for the treatment of electric shock at least in every sub-station.

Particular attention is drawn to the following regulations:

Regulation 22

'Adequate precautions shall be taken to prevent any *conductor* or *apparatus* from being accidentally or inadvertently electrically charged when persons are working thereon'.

Regulation 28

'No person except an *authorised person* or a competent person acting under his immediate supervision shall undertake any work where technical knowledge or experience is required in order adequately to avoid *danger*; and no person shall work alone in any case in which the Secretary of State directs that he shall not. No person except an *authorised person*, or a competent person over 21 years of age acting under his immediate supervision, shall undertake any repair, alteration, extension, cleaning, or such work where technical knowledge or experience is required in order to avoid *danger*, and no one shall do such work unaccompanied.

'Where a contractor is employed, and the *danger* to be avoided is under his control, the contractor shall appoint the *authorised person*, but if the *danger* to be avoided is under the control of the occupier, the occupier shall appoint the *authorised person*'.

Regulation 31

'Every *substation* shall be under the control of an *authorised person*, and none but an *authorised person* or a person acting under his immediate supervision shall enter any part thereof where there may be *danger*'.

A set of 'works rules' should be drawn up covering the operation of the high-voltage system and the precautions that have to be taken when work has to be done.

A simple set of rules as shown in Fig. 21 should be regarded as a minimum requirement.

It should be noted that, to comply with Regulations 28 and 31, the individual who is responsible for the operation of the high-voltage system is named, together with his deputy to take care of absence due to illness or holidays. It is essential that those so named are fully competent to carry out the duties that they have to perform.

For larger installations, and particularly those where operation is directed from a control room, a detailed rule book should be prepared to suit the circumstances, so that all personnel are acquainted with their individual responsibilities and the operating procedures for the works in general.

A copy of the works rules should be given to all personnel involved or likely to be involved. Further, whenever contractors work on the site, a copy should be given to them so that they have no excuse for not complying. Fig. 21 is a simple example.

15.2 Labels and notices

To avoid misunderstandings when issuing a permit for work to be carried out, it is important that a comprehensive system of notices and labels should be applied.

The following must be clearly identified:

- substations
- circuits
- other items of equipment, e.g. transformers.

Conspicuous notices should be provided outside each substation, stating

- the name of the substation
- the voltage
- warning against unauthorised entry
- how to raise the alarm in case of fire.

XY COMPANY: Z WORKS

REGULATIONS FOR WORK ON HIGH-VOLTAGE ELECTRICAL EQUIPMENT

1 Scope

These Regulations apply to *all*

high-voltage (i.e. exceeding 650 V) apparatus.

high-voltage cables and connections.

other equipment of *any* voltage in substations containing high-voltage apparatus.

2 Statutory regulations

The Electricity (Factories Act) Special Regulations 1908 and 1944 apply.
A copy is posted (e.g. on the works notice board).

3 Operation of high-voltage system

The person authorised by the company to operate the high-voltage system is

Mr. ...., Works Electrical Engineer

Works telephone number ....

Home telephone number ....

Home address ....

....

....

His authorised deputy is

Mr. ...., Electrical Foreman

Works telephone number ....

Home telephone number ....

Home address ....

....

....

No other person shall operate or interfere with any high-voltage electrical apparatus or associated equipment unless specifically authorised in writing.

4 Permit to work

No work shall be carried out unless a written permit on the company's official form has been issued by the Works Electrical Engineer or his authorised deputy.

Fig. 21

- how to send for the ambulance
- who keeps the key.

Similarly, conspicuous notices should be provided inside each substation, stating

- electric-shock instructions
- warning against working without a permit
- circuit labels (on the back and front of gear where applicable)
- apparatus identification labels (e.g. on transformers)
- single-line diagram
- any applicable routine instructions.

Circuit labelling must be explicit, but short. Various systems are available. Continual repetition of the same numerals, e.g. 'no. 1 and no. 2 substations', 'no. 1 and no. 2 feeders', 'no. 1 and no. 2 transformers', can be confusing. Names, rather than numbers, are better for substations, e.g. 'ABC plant'. This, in turn, means that the associated feeders and interconnections will bear names, but, in addition, a consecutive numbering sequence is helpful. Thus, in addition to the name of the substation at the far end of the feeder, the labels at both ends will bear a common numeral.

Similarly, a separate series of numbers can be reserved for transformers.

Where the switching is directed from a control room, it may be convenient for each circuit breaker to bear an identification symbol comprising a letter, by way of a prefix denoting the location, together with, perhaps, three digits indicating voltage and duty.

As far as circumstances permit, any work to be undertaken should be properly planned in advance. This should include checking that all gear and materials are available. Although unforeseen difficulties may be encountered, the need to improvise should be avoided where possible. This will prevent delays and the consequent temptation to take risks.

The implications of switching 'off' must be considered. Apart from interrupting production, someone may be stranded in a lift or in a dangerous position because the lights have gone out, flooding may occur because a drainage pump has lost its power supply, and so on.

Before the actual work is begun, the equipment itself must be made dead and steps taken to ensure that it will not be made alive again until the work is finished. The equipment to be worked on must also be connected to earth, and, furthermore, precautions must be taken to prevent those doing the work from touching other plant that is still alive—whether by inadvertence, curiosity or overenthusiasm. Similarly, when re-energising after the work has been done, it is essential that all men be withdrawn and that none of them returns 'just to finish something off'.

Although, in simple circumstances, it might appear practicable to carry out these precautions by word of mouth, there are two reasons against this. First, the least complication introduces risks that cannot be justified. Secondly, carrying out work as a result of verbal instructions in one instance makes it difficult to enforce the full precautions in another. Therefore, a safety procedure involving a written 'permit to work' should apply on *every* occasion when work has to be done.

While extensions, or new work, can usually be installed in safety up to the point where they have to be connected to a supply, a permit should still be issued for carrying out the work. The recipient of the permit then automatically becomes responsible for his men and for withdrawing them when the time comes for cancelling the permit prior to energising the new work.

Provision should be made for recording all switching operations, operation of protective gear, and work done. Nothing elaborate is needed, but, when a permit to work is issued, a crossreference in the log book should quote the number.

Where convenient, one log will suffice for a works h.v. distribution system, but an alternative is to have a book in each substation for an entry to be made at each visit.

The most important safety precaution is that a permit to work should be issued before any work is done. A general form of permit is shown in Fig. 22. It should be noted that this amounts to a recital of the precautions that have been taken in any instance.

The permit forms should be contained in a duplicate book and numbered sequentially for easy identification. It is convenient for the top copy to be on ordinary paper and to remain in the book, and for the duplicate, which is torn out and given to the recipient, to be on thin card, which is less easily damaged or lost.

In general, only one permit book must be in use in any works at one time, although it may be necessary to subdivide very large installations and have a separate book for each well defined sector. Cancelled permits should be filed, and, when the permit book is exhausted, it too should be filed for reference when a new book is brought into use.

If, for any reason, a permit is partly written out and then not issued, it should be boldly endorsed 'NOT USED' on both the original and duplicate and the duplicate filed in the usual way.

15.3 Making apparatus safe for work

To comply with Regulation 22, the general procedure is as follows:

- switch off
- isolate
- lock off
- earth
- post 'danger' notices
- issue permit.

Before switching off, a check should be made that those who will be deprived of a supply will not be endangered.

XY COMPANY: Z WORKS

PERMIT TO WORK ON HIGH-VOLTAGE ELECTRICAL PLANT AND ASSOCIATED APPARATUS

Location .... Voltage ....

Exact description of equipment on which work has to be done ....
....
....

Brief description of work to be done ....
....

Exact description of how the above equipment has been isolated and locked off from the supply ....
....
....

The equipment to be worked on has been earthed at the following points, and 'danger' notices posted ....
....
....

Automatic fire-protection equipment has been locked off ....
....

ALL OTHER EQUIPMENT IS DANGEROUS

Issue

I CERTIFY that the above safety precautions have been carried out and that the equipment listed is dead. It will *NOT* be made alive until a clearance is given by the recipient of this permit.

(signed) .... (date) .... (time) ....

Receipt

I ACCEPT RESPONSIBILITY for carrying out work on the equipment listed in this permit. No attempt will be made by me, or by any man under my control, to work on any other equipment.

(signed) .... (date) .... (time) ....

Clearance

I DECLARE that the work for which this permit was issued has been completed and that the equipment is ready for service. All men under my control have been withdrawn and warned that it is no longer safe to work on the equipment.

(signed) .... (date) .... (time) ....

Cancellation

This permit to work is hereby cancelled.

(signed) .... (date) .... (time) ....

Fig. 22

After switching off, the circuit concerned must be isolated. With parallel or ring-main circuits, this applies to both ends and, for teed circuits, all ends. If there is a potential transformer, this too should be isolated on both the h.v. and l.v. sides.

Any access to live conductors should be padlocked. Usually, only the shutters to the busbar spouts are involved, but this is not necessarily so. It may also apply to the shutters of an incoming feeder or interconnector that has to be left alive, and, where applicable, the shutters guarding the spouts of potential transformers should also be padlocked.

Most modern switchgear incorporates (or is provided with adaptors for) earthing facilities. These should always be used. With metalclad gear, it is better to earth immediately after making sure that all is safely isolated, without resorting to any potential testing device, which involves access to the spouts. Trying to get into the spouts is an awkward business, and can be dangerous if a mistake has been made and the circuit is still alive. Proper earthing equipment should discharge the conductors safely, even if the consequences may be an interruption in the supply.

Similar considerations apply with ring-main isolators, except that the operations of switching off and isolation are combined. These isolators are easily (perhaps almost too easily) moved to the earth position. Modern isolators are spring-assisted, so that they can be closed in safety onto a short circuit if this should happen inadvertently. Older designs may explode if maltreated in this way.

In any event, great care should be taken to ensure that, by proper isolation, the circuit is dead before attempting to earth it.

In other circumstances, a potential testing device should be used, but this must itself be tested before *and* after use. This applies where it is necessary to open up apparatus, such as a transformer, and on old switchboards (e.g. cell-type gear) without proper earthing facilities. After the circuit has been checked to make sure that it is dead, it should be immediately earthed by first hooking on substantial test leads, handled with an insulated pole, after which the earth connections should be bolted on.

The use of flexibles as improvised earthing equipment for 'flashing' conductors is a dangerous procedure, and must be avoided.

If work has to be done on a cable, the circuit concerned should be earthed at the switchgear, and at both ends where there is any possibility of a back feed.

Unless the work is in the immediate vicinity of the switchgear, so that the cable can be traced beyond any shadow of doubt, a cable-spiking device should also be used at the point where the work is to be done. This requirement applies irrespective of the voltage.

When the circuit breakers are withdrawn, so that they are physically separated from the live metal of the switchboard, earthing is unnecessary, but, as has already been emphasised, the shutters to the spouts must be padlocked.

A danger notice should be hung on the switchgear where earthing has been carried out (at both ends where necessary). If shift working is done and the permit may have to be cancelled by another shift, the danger notices should have a space in which the permit number can be entered, as a guide to helping the relief find the appropriate permit.

Sometimes danger notices are marked 'only to be removed by Mr. . . .'. While such an arrangement may be convenient for simple medium-voltage circuits, it is not acceptable as an alternative to issuing a permit where h.v. equipment is involved.

One other precaution may have to be taken; any automatic fire-protection equipment, e.g. CO_2, must be locked off in the zone where work is carried out. If the work, say cable jointing, is being done outside the substation from which the circuit is controlled, the substation fire-protection equipment should be left operative. It must be remembered that a gas such as CO_2 is heavier than air; so the substation equipment must be locked off if there is any risk of it 'flowing' into a cable basement or tunnel where the work is being done. Unfortunately, while work is being carried out, the risk of fire will be greater than normal; e.g. it may be caused by a mistake or by the gas from a jointer's blowlamp. It will then be necessary to have someone available to operate the fire-protection equipment by hand control as soon as all personnel have been withdrawn. In some instances, the man in charge could be given authority.

When all has been made safe for the work to be done, a 'permit to work' *must* be issued before work is started. No work involving the use of tools should be carried out in any substation containing high-voltage apparatus until the permit to work has been completed. There must be no exceptions to this rule.

That the work does not concern high-voltage gear is immaterial; it must be clear to the recipient of the permit just what work he is allowed to do. Although only the lighting installation or the medium-voltage switchgear may be involved, if a permit is issued, there is no excuse for the recipient misunderstanding his instructions and working elsewhere, with possibly dangerous consequences.

Every permit must state clearly where the work has to be done, what it entails and what safety precautions have been taken. The permit form may not exactly suit every kind of work that may be encountered. If some of the spaces on it are not applicable, it should be stated *why* they are not applicable, and on no account should any blanks be left.

Descriptions such as 'no. 2 transformer o.c.b.' may give the impression that the whole o.c.b. panel is safe, whereas the busbar spouts, even though padlocked, are most likely to be alive. If the circuit breaker has been withdrawn from the gear, the current-transformer chamber may not have been earthed if it was not intended that any work had to be done on it.

When the permit form has been carefully completed, it must be signed both by the authorised person issuing it and by the recipient, who will keep the duplicate. The recipient of a permit to work should be encouraged to watch the safety precautions that are carried out, and there is no objection to him fitting his own padlocks, by way of an additional safeguard, at the relevant points.

Apart from the issue of the permit, the authorised person concerned must be satisfied that the recipient is competent to carry out the work which the permit allows him to do. The recipient must take the responsibility for his assistants, and it should be noted that, under Regulation 28, he may not work unaccompanied.

Wherever possible, only one permit at a time should be issued for any circuit. Sometimes this is not practicable, as it may be necessary, say, to issue one permit for work on a cable and another for work on a transformer on the same circuit. There is the obvious danger that it would be fairly easy to cancel one permit and forget the other. In such an instance, the two permits should be crossreferenced and two 'danger' notices hung on the switchgear controlling the circuit, so that permit numbers are exhibited separately.

The effectiveness of safety precautions is often affected by the nature of the work that is being done.

For instance, a cable may have to be cut. Even if one end of the circuit is earthed and the cable has been earthed again by spiking, the other end may be left unearthed after the cutting. This may not matter on a dead-end circuit, where there is no risk of a backfeed, but it would obviously be dangerous with overhead lines, which could be energised by induction or static.

Changing a current transformer is a similar case. Care must be taken to avoid this kind of danger, but circumstances are so varied that each instance must be judged on its merits.

During the course of the work, it may be necessary to remove, or partly remove, earth connections to test the connections with a very low-voltage instrument, such as a bell set. This would be legitimate, provided that the locks on the busbar shutters do not have to be disturbed and that the earthing is restored as soon as the tests are completed. Such testing should, however, be done under the immediate supervision of the authorised person who issued the permit.

If the circuit has to be made alive temporarily for phasing out at full voltage using 'test sticks', the permit must be cancelled. The testing would then be carried out by the authorised person, or under his immediate supervision. Subsequently, another permit for the completion of the work would be issued when it was safe to do so.

Safety precautions, i.e. isolation, earthing, padlocking etc., must not under any circumstances be disturbed until the recipient has cancelled the permit. When the permit is cancelled, it must be replaced in the duplicate book and signed by the recipient, so that the signature appears on both the top copy and the duplicate. It is best if the authorised person, who issued the permit, then countersigns it; but, with shift working, this is not always practical.

It is not admissible for a deputy to the recipient of a permit to sign the cancellation. If the individual has left the site, he must be brought back—however inconvenient it may be. It will be apparent, therefore, that, if there is a possibility of a circuit for which a permit has been issued having to be made alive in an emergency in the middle of the night (assuming that the state of the work allows this), the permit should be cancelled before he leaves in the evening and a new permit issued next morning.

Whereas the permit procedure should be scrupulously followed, unforeseen circumstances do arise; the recipient of the permit may have lost it. If this should be so, he must sign the top copy in the permit book, noting specially that he has lost the duplicate and undertaking to consider it as cancelled if it should ever be found.

Another possibility is that the recipient of a permit may be taken ill and be incapable of cancelling it. It would then be best for the authorised person to explain the position to the works manager, and, after saying what precautions have been taken to prevent the recipient resuming work when he returns, obtain special sanction for overriding the works rules.

Where a factory is fed from the supply authority's mains by two or more circuits in parallel, or where the consumer has generating plant that could feed back into the supply, the procedure for issuing permits should be discussed with the supply authority and the procedure clearly stated in the works rules, so that the responsibilities of both parties are defined.

In a small installation, there may be only one man authorised to work on high-voltage gear. If he has to do so, he should write himself a permit; this will at least act as a reminder, and provide a record that safety precautions have been taken. Nevertheless, he must have a reliable mate, as, according to Regulation 28, he should not work unaccompanied. As far as practicable, his mate should be trained to take the correct action should an emergency arise.

In some small installations, there may be no one who has the necessary experience to be competent to work on high-voltage equipment, and, also, the necessary equipment may not be available. Where this is so, a maintenance contract should be arranged. Often the local electricity board is willing to undertake this work, and will have its own safety code, as also will a reliable maintenance contractor.

A permit to work should always be issued when any work has to be done on a high-voltage motor, just as with any other high-voltage apparatus.

Sometimes work has to be done on driven plant, where fortuitous starting could be extremely dangerous. This situation often applies in steel-rolling mills, where the electrical equipment is located in a separate motor room.

In such circumstances, the authorised person (who would normally issue the permit for electrical work to be done) should also issue the permit to the fitter carrying out work on the plant driven by the motor. If no electrical work has to be done, earthing the circuit is not strictly necessary, but it is probably better to make a routine of it, so that it is not omitted when it is necessary.

It should be particularly noted that this danger with respect to the driven plant could apply irrespective of the voltage of the motor, and, where necessary, the works rules should be extended accordingly. This is another instance where it may not always be easy to lay down the line of demarcation, as it would obviously be impractical to apply a permit system to drives such as machine tools.

Special authority should be given to the cleaners responsible for any substation containing high-voltage equipment. This should be in writing, naming the individual in charge and stating clearly what apparatus must not be touched. Again, it should be noted that, under Regulation 28, even a cleaner may not work unaccompanied.

The nature of the work that has been done will dictate what testing should be carried out. Although, in general, a pressure test should be made on new high-voltage connections, it must be remembered that pressure testing overstresses the insulation. Therefore, a minimum of such testing should be done; if old insulation must be subjected to test conditions because it is joined to new, the severity of the test should be relaxed with respect to both time and pressure, compared with what would normally be applied to a wholly new insulation.

For testing, under Exemption 4, the Electricity Regulations do not apply; instead, special precautions have to be taken to ensure safety during this operation.

A separate permit should be issued for carrying out pressure tests, and this should state that the circuit earthing will be removed prior to the test. If practical, the circuit earthing should not be disturbed until after the temporary test connections have been attached.

Access to any other conductors that have to remain alive while testing is in progress must remain locked off.

During testing, great care is essential. None of the normal safeguards will be operative, and the voltage will be in excess of normal and, therefore, more lethal. Provision must be made, and used, for discharging the conductors at the conclusion of each test.

After testing has been satisfactorily completed and all permits cancelled, 'danger' notices should be removed. If any earths remain on the circuit, these too should be removed before the circuit breaker or breakers are returned to the service position. Padlocks or other locking devices may be removed, but only as is necessary as this operation proceeds.

Before switching on, a check should be made that all control circuit fuses and trip links are in place, and a second check should be made to ensure that no earths have inadvertently been left on the affected circuit.

15.4 Conclusion

Carrying out the routine precautions leading up to the issue of a 'permit to work' on high-voltage conductors, together with its subsequent cancellation, is a nuisance and time-consuming. There will always be a temptation to take short cuts. Nevertheless, if accidents are to be avoided, meticulous discipline is essential to ensure that the rules are conscientiously obeyed.

16 SUBSTATION DESIGN AND LAYOUT

Buildings for substations should, as far as possible, be resistant to fire and explosion. This requirement must be considered both with respect to anything happening inside and with respect to anything extraneous.

Where oil circuit breakers etc. are used, there is a remote risk that oil may be vaporised and eventually detonated. If this happens, it is the building that is likely to suffer more than the equipment. The effects of blast are fickle, but some provision for pressure relief should be included wherever it is reasonably practicable. For instance, windows may serve this purpose and, at the same time, avoid complete dependence on artificial light. There should be nothing to be ashamed of inside a substation, and windows may deter illicit 'recreation' during the night shift; furthermore, if there should be a fire, it may be noticed earlier. However, it is not in every locality that the integrity of the glass can be maintained.

Weatherproof construction is obviously essential, and so is convenient and easy access. Whereas the doors must be big enough to get the gear in, it is only for exceptionally heavy equipment that double doors are needed. The latter are normally difficult to make weatherproof, and are less convenient than a single door for both normal and emergency access.

There is no absolute legal requirement demanding that each substation should be provided with two doors, but this is a sensible precaution for all but the smallest switchrooms. The layout should be such that there is always an alternative escape route if one is blocked. Furthermore, the alternative route should be available to reach anyone who has been disabled as a result of an accident. For these reasons, each door should always be easily openable from the inside without a key, i.e. with a panic latch, and from the outside with a key. Each door should open outwards, and there should be no hazards immediately outside, e.g. steps to fall down, obstructions to trip over or passing traffic from which anyone escaping might suffer harm.

Space for access to switchgear should not be skimped. Manufacturers' drawings usually give the absolute minimum; nearly always, a little more is desirable. For instance, if a circuit-breaker truck is withdrawn for maintenance, there must still be room to get past. If this is not possible, a valuable escape route may be blocked in an emergency.

A small substation might comprise only a single room sufficiently large to house an h.v. ring-main unit and an m.v. switchboard with the transformer outside. Provision for cables by means of trenches over which the switchgear can be placed and ducts for bringing in the cables will be needed in the foundations.

There will not be much to burn inside a simple substation such as this, and hand fire extinguishers of the CO_2 or BCF type will suffice. If burning oil escapes from the transformer and is likely to affect anything else, there should be a sump to trap it and prevent further spread of fire until the fire brigade can tackle the blaze. It must be stressed that fires in installations of this category are extremely rare, and, whereas nothing must be done to encourage the spread of fire, it is hard to justify any great expenditure on elaborate precautions. For instance, if a heavy and persistent short circuit should set fire to the ring-main unit, which is the only apparatus containing oil, the latter will be so severely damaged that it will need replacement anyway. The same can be said of the transformer outside.

Larger installations require more careful thought. On a long switchboard, a conflagration at one point must be extinguished before it spreads and eventually destroys the whole board. Fixed firefighting equipment, such as the CO_2 or BCF type, will be needed; in addition, if there is a bus-section switch, it is usual to provide fire walls on each side of it. The fixed firefighting equipment must protect cable trenches and basements, where experience has shown that burning oil or compound from a switchgear fire may cause more damage to the cables than to the switchgear installation. In addition, any holes through floors should be sealed.

Various sections of the equipment should be located in separate compartments; e.g. each switchboard should be in a different room, and there should be a separate control or relay room, a battery room and accommodation for ancillary equipment such as battery chargers, gas cylinders for fire protection etc. The transformers outside should be separated by fireproof walls.

While it may not always be easy to draw the line and decide where fixed firefighting equipment is necessary, precautions of the type described above will only be justifiable on the larger installations. A reasonably liberal provision of hand extinguishers should always be provided. Transformers outside can be considered as a job for the fire brigade, though there may be exceptions to this if there are any other fire risks in the immediate vicinity.

In the vicinity of switchgear, some thought must be given to cable runs that would be vulnerable to burning oil. A few cables in the bottom of a trench can be covered with sand if their positions are recorded, but cables on hangers are less easy to protect and, except where fixed fire protection is installed, should be avoided where there is the risk of an oil fire. By careful layout, it is often possible to avoid a concentration of important cables.

The equipment in substations must be kept dry, but there is no particular merit in keeping it warm. Thus dehumidification equipment may justify its higher first cost, because it is cheaper to run than the more usual electric heaters.

Good artificial light is always a necessity. Some emergency lighting is a great advantage when dealing with a breakdown. For this purpose, one or more self-contained automatic sealed battery units with integral chargers are suitable. A tripping battery should not be used for the purpose, owing to the risk of it being accidentally discharged by the lighting.

17 CONCLUSION

It will be seen that the design of an industrial electricity-distribution system involves many different aspects of electrical engineering that must not only be co-related between themselves but also to the problems of the plant or process that the distribution system serves.

Although any particular problem can usually be solved in isolation, it will be ineffective to do this unless the final result fits in with the overall pattern of the system.

In conclusion, the author acknowledges with thanks the ready co-operation that he has received from his colleagues in McLellan & Partners and the many branches of the electrical manufacturing and supply industries.

18 INDEX

alternator, *see* generator

cables (*including* feeders)
general, **2–3, 12–18, 27, 36, 40, 44, 48–49, 51, 56–57, 62, 68–70, 79–81, 85, 87–88**
runs/routes, **11, 68–70, 77, 79, 88**
circuit breakers, **15, 18, 21–22, 24–25, 27–28, 32, 35–36, 39, 41, 43–45, 49, 54, 66, 72, 79–80, 83, 87**

direct current, **20, 25**
demand, **1, 3, 60, 62, 74**
discrimination, **14, 19, 29, 41, 42–45, 60**
distribution systems
general, **1, 7, 9, 11–15, 18–21, 25–26, 30, 45, 58, 62, 72, 74, 78, 90**
radial, **11, 14, 16**
ring-main, **16–18, 27, 88**
feeder–transformer, **16–18, 66, 68, 77**
diversity, **1–3, 11**

earthing
general, **40, 47–53, 56, 77–84, 86**
system, **40, 50–55, 60**

faults (*including* fault currents), *see* short circuits
feeders, *see* cables
fire precautions, **24–25, 69, 76, 80, 87–88**
frequency variations, **26, 30–31, 73**
fuses, **14–15, 19, 22–23, 25, 27–28, 35, 39, 40–41, 43–45, 49, 84**

generators (*including* alternators), **24, 31–33, 48, 53–55, 57–63, 73, 84**

harmonics, **20, 54, 65–66**

instrumentation, **33, 71–74, 83**
interruptions, **14, 16, 26–27, 29, 41, 49, 60, 78, 81**

load
general, **1–5, 11, 13–17, 20, 22, 24, 29–31, 35–39, 41, 53, 55, 57–60, 63–66, 70–74**
assessment/growth, **1–3, 5, 26, 39**
installed, **1–2, 40**
factor, **1, 13**
losses, **12, 16, 27–28, 44, 62, 66, 74**

overhead busbar systems, **15**

permits to work, *see* safety
power factor, **23, 30–31, 59, 62–67, 69, 72–73**
protection, **14–15, 20, 23, 25, 28–29, 40–42, 44–45, 49–51, 53–55, 59–60, 68–69, 71, 79**

reliability, **11, 16, 26, 28, 31, 33**
rates
local authority, **7–9**

safety (*including* permits to work), **75–86**
short circuits (*including* faults *and* fault currents), **11, 14–18, 21–24, 27–31, 35, 39–57, 59–60, 63, 70–71, 81, 88**
substations
design and layout, **87**
general, **2, 5, 7–8, 10–11, 14, 16–18, 25, 27, 35–36, 41, 50, 65–66, 68, 70–71, 75–79, 82, 85, 88**
supplies
incoming, **1, 2, 5, 7–8, 17, 28, 31–33, 38, 43, 45, 50, 53–55, 58–60, 63**
general, **14, 16, 18–20, 26–32, 36–37, 40–41, 43–44, 48–49, 51, 53–55, 72–74, 79–80, 84**

switchgear, **2, 11, 14, 16, 18, 21–22, 24–25, 31, 34–35, 39, 44, 46, 56–57, 59, 63, 66, 71, 81–83, 87–88**

transformers, **2, 3, 7–9, 11–12, 14, 16–21, 23–25, 29, 35–44, 48–51, 53–57, 59, 62, 65–66, 68, 71–74, 76, 78, 81–83, 87–88**

vital services, **26, 31–32**
voltage
  choice, **5, 8, 11, 12, 15, 18–21**
  dips, **26, 28–29, 63**
  irregularities, **20, 26, 30, 59, 63**